画意中的建筑

彭一刚手绘建筑表现图集

彭一刚 著

华中科技大学出版社
http://www.hustp.com
中国·武汉

编者按

这是一部建筑巨匠的手绘力作,历时40余载集结成册;这是一部匠心之作,彭一刚院士手绘首次系统整理出版。精湛丰富的绘画技法和独具匠心的表现,让一栋栋建筑在大师笔下出神入化、令人震撼。独家收录300多幅精美绝伦的建筑手绘图,让读者尽享手绘盛宴。彭院士的手稿堪称旷世大作,他将建筑手绘引领到一个全新的高度。他的作品让无数建筑师顶礼膜拜、惊叹赞赏,为无数建筑学子提供可以学习、临摹的卓越范本。本书是建筑师、建筑院校师生重要参考用书,值得学习和珍藏!

图书在版编目(CIP)数据

画意中的建筑:彭一刚手绘建筑表现图集/彭一刚著. —武汉:华中科技大学出版社,2018.5
(哲匠书系)
ISBN 978-7-5680-3588-0

Ⅰ.①画… Ⅱ.①彭… Ⅲ.①建筑画—钢笔画—作品集—中国—现代 Ⅳ.① TU204.132

中国版本图书馆CIP数据核字(2018)第065918号

画意中的建筑:彭一刚手绘建筑表现图集
HUAYI ZHONG DE JIANZHU: PENG YIGANG SHOUHUI JIANZHU BIAOXIAN TUJI

彭一刚 著

出版发行:华中科技大学出版社(中国·武汉)	电话:(027)81321913
地　　址:武汉市东湖新技术开发区华工科技园	邮编:430223

特约指导:吴家骅
责任编辑:张淑梅
责任监印:朱　玢

印　　刷:武汉精一佳印刷有限公司
开　　本:787 mm×1092 mm　1/12
印　　张:24.5
字　　数:147千字
版　　次:2018年5月第1版　第1次印刷
定　　价:139.00元

投稿邮箱:zhangsm@hustp.com
本书若有印装质量问题,请向出版社营销中心调换
全国免费服务热线:400-6679-118　竭诚为您服务
版权所有　侵权必究

写在前面

还是在新世纪之初，公差深圳时，受到《世界建筑导报》杂志社吴家骅社长的热情接待，并示以由他社出版的《陈世民草图集》，长达800余页，印制十分精美。在羡慕之余，曾探询可否为我出版作品集，立即得到他的欣然允诺。之后不久便为我出版了《画意中的建筑——彭一刚建筑表现图选集》和《传统·现代·融合——彭一刚建筑设计作品集》两部作品集，印制都达到了非常高的水平，我非常满意。

但当时我忽略了一个问题，即杂志社没有发放书号的权利，幸好吴社长神通广大，以一家香港出版社的名义出版发行，甘冒亏本的风险，可见吴先生为友情不惜两肋插刀。然而，这条渠道无法与内地销售系统接轨，致使在国内销售不畅，据说还有少量书压库，这自然会影响他们的经济效益。

此时，华中科技大学出版社建筑分社的张淑梅编辑来到我家，说是看过这两本书，很实用，会受到读者欢迎，但在网上无法查到，颇感遗憾。我当时表明愿改由华中科技大学出版社出版发行。可是，总不能一稿两投吧。于是，我声明：若变更，只能由你们和吴社长协商，并一再强调切不可损伤吴总他们的利益。时隔不久，得知已经获得吴社长的同意，于是问题顺利解决。

我的这两部作品集，比较全面地收集了我的建筑表现图和设计创作作品，虽然如鸡肋，似敝屣，但有人愿意出版，我还是愿意把它奉献给更多的读者。值此出版发行之际，本书增加了后续篇章节，以飨读者。

彭一刚

目录 / CONTENTS

前　言 / PREFACE　　　　　　　　　　　　　　　1

1　蓝色篇 / BLUE　　　　　　　　　　　　　　　3

2　棕色篇 / BROWN　　　　　　　　　　　　　　35

3　黑白篇 / BLACK & WHITE　　　　　　　　　117

4　彩色篇 / COLORS　　　　　　　　　　　　　139

5　草图篇 / SKETCHES　　　　　　　　　　　　193

6　其他篇 / MISCELLANEOUS　　　　　　　　　223

7　后续篇 / SUPPLEMENT　　　　　　　　　　　261

[附]　传统与时代圆满的契合　　　　　　　　　283
　　　回顾徐中先生的代表作——原外贸部办公楼

前 言

"诗情画意"是艺术家所追求的一种高雅境界。诗情,我不敢高攀,因为它似乎要求太高了。但画意却是我所要追求的。那么,在"画"字后面缀上一个"意"字是什么意思呢?我认为意就是一种意念,是属于主观的范畴,更明白地说,就是在作画时要把作者的主观追求渗入到画中去,而不是单纯地、如实地描绘对象,这样,就赋予画以艺术意蕴。建筑师的表现图重在"表现"而非"再现",这之间的差别往往体现出建筑师的艺术修养和素质的高低。于是,问题就出来了:电脑绘制的表现图该如何定位呢?单就电脑本身来讲其功能只限于再现,并且绝对准确、真实、可靠,但在不同人的运用中,还是可以赋予它以不同程度的"表现"成分,从这个意义上讲,它也可以纳入到艺术的行列。然而,这种"赋予"却是有限的,其原因就在于它太真实了。我的表现图则并非绝对真实,而在"似与不似"之间,只是让人看不出破绽而已。要是与狂放的国画大师相比,还是"收敛"得多,不敢"任意胡为",否则就让业主不信任了。

这本集子所收录的是近 40 年来从事实际工程的方案设计,其中,有的已经建成,但相当一部分由于各种原因没有建成。从表现图的角度看是否建成无关紧要,故一律隐去工程的名称。由于作画的时间段和形式不同,这本画册可以分为以下篇章:蓝色篇、棕色篇、黑白篇、彩色篇、草图篇、其他篇以及后续篇等 7 个篇章,每一篇章都有一个简短的说明。

彭一刚

Preface

Artists always pursue elegant realm which is being poetic and artistic. However it's too high demanded for me to pursue poetic. But still I have to be artistic with my works. For that purpose, painters have to give their subjective pursuing to the works rather than to faithfully describe the object. Architects' drawings emphasize representation rather than reproduction. Differences between the two processes can show an architect's cultivation. Nowadays, people tend to use computers to draw their drafts. And those e-drawings are perfectly real, accurate and dependable. But just because it's too real, the limitation of being artistic is added. Compared to the drawing of computer edition, my presentation drawings are not that real. However, people can't realize the unreality from my drawings. Why? Because the nature of my drawings is the realm between likeness and unlikeness — it's the artistic conception!

All the project proposals in this book are the real ones since our nation's reform and opening. Some of them were already constructed while some were not, but it has nothing to do with the presentation drawings. Therefore all the project names are faded. This book has seven parts separately named as BLUE, BROWN, BLACK & WHITE, COLORS, SKETCHES, MISCELLANEOUS and SUPPLEMENT which are categorized by time and type.

Peng Yigang

1

蓝色篇

　　以前作画工具主要是小钢笔和鸭嘴笔。改革开放初期引进了意大利生产的"针管笔"，比起前者要好用得多。但随笔所赠的墨水只有一瓶，并且为黑色，用来作画似感沉闷。后来发现改用天津生产的鸵鸟牌蓝黑墨水效果也相当不错，于是在一段时间我便用它来作画。鸵鸟牌墨水虽属一般的书写墨水，但色泽深沉且无渣滓，并且也适用于渲染。既适合作钢笔表现图又适合用来渲染，于是我的一部分表现图就是以渲染和线条结合的方法完成的。在这个集子中我选用了其中的一部分作为那个时期的代表作。

BLUE

Years and years ago, we only had fountain pen and border pen as our tools. At the beginning of our nation's reform and opening, we applied needle pen which was made by Italy. It's much better than the former ones, but there's only one bottle of attached ink for each pen, and the color is black. I felt dull with the ink after some time. Later I found the dark blue ink which was made in Tianjin with a famous OSTRICH BRAND. And within a period of time I drew with this ink and got a wonderful effect. Ostrich was a common ink , but its deep color was very good for coloring. Naturally, a part of my presentation drawings was completed by the Ostrich Ink which brought me a combination of blue lines and coloring. I chose some drawings from that time for this book.

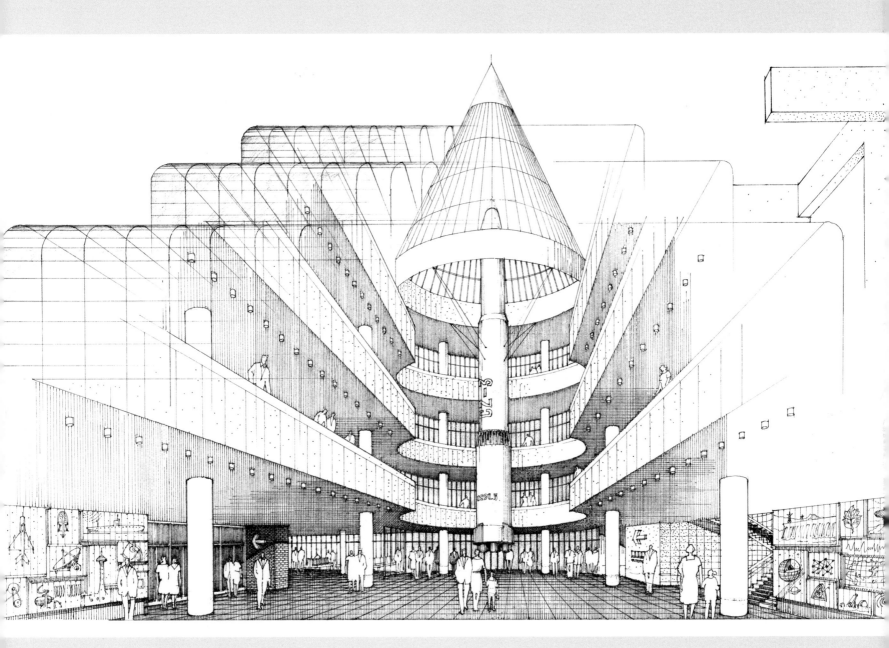

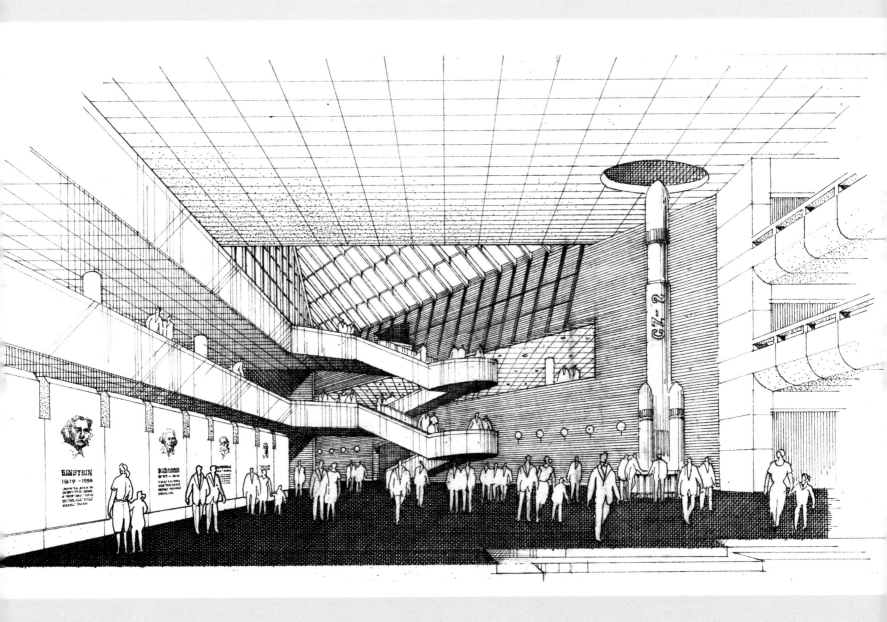

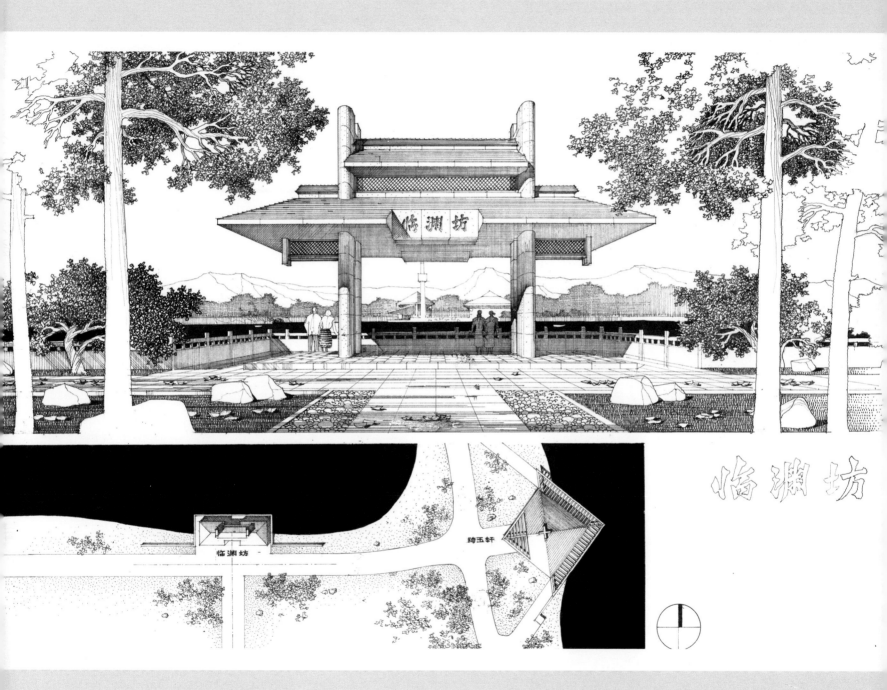

此方案系天津大学聂兰生教授所作

绮玉轩

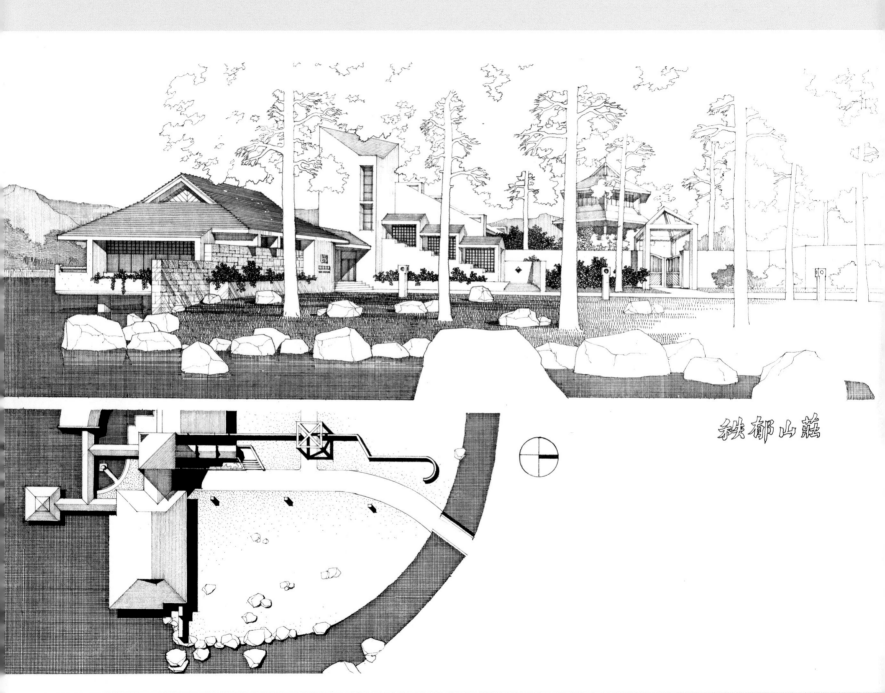

秩郁山荘

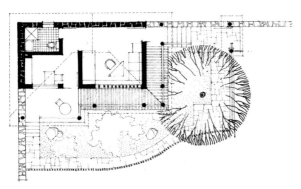

24

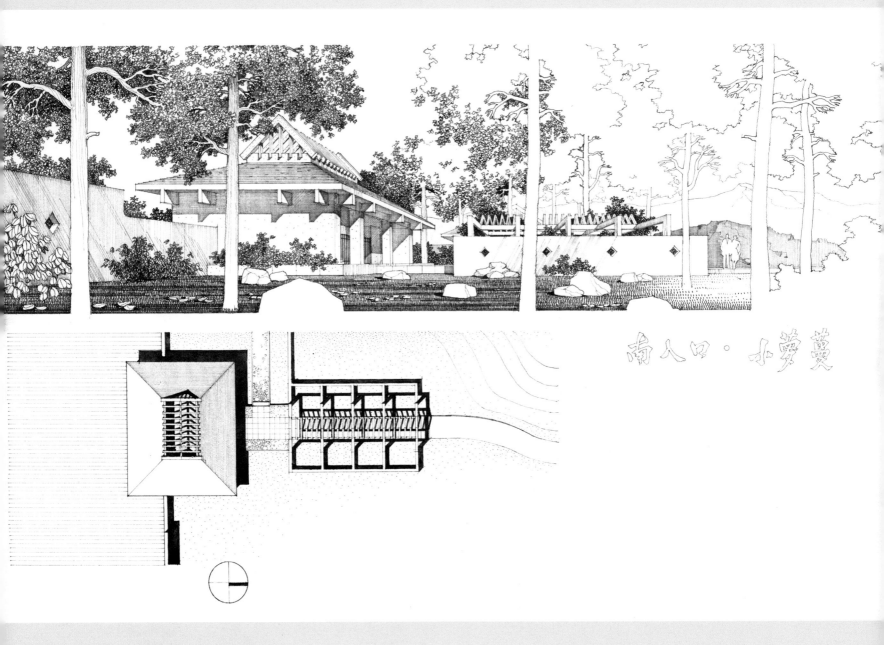

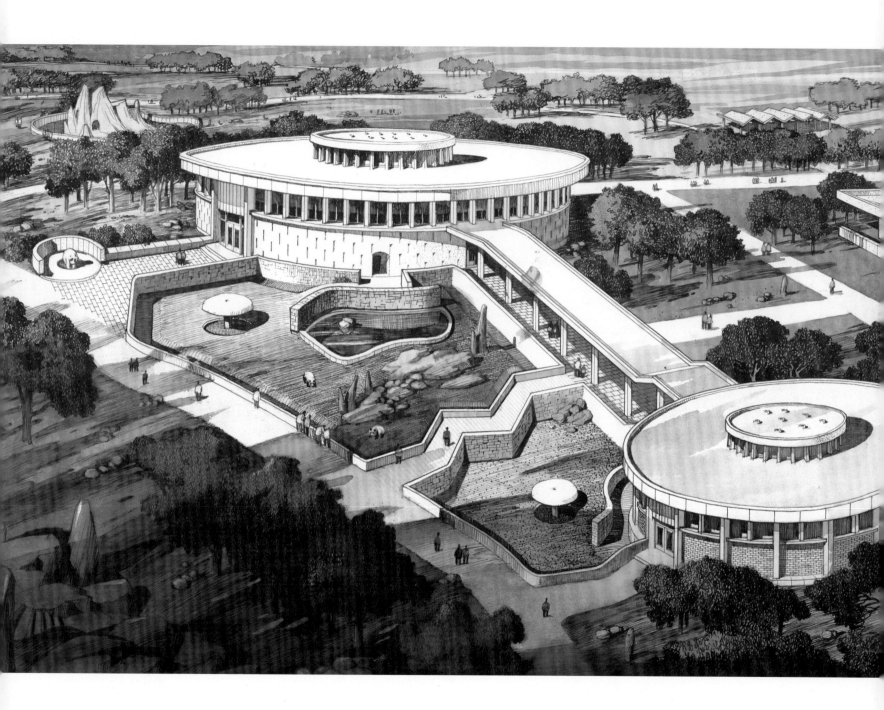

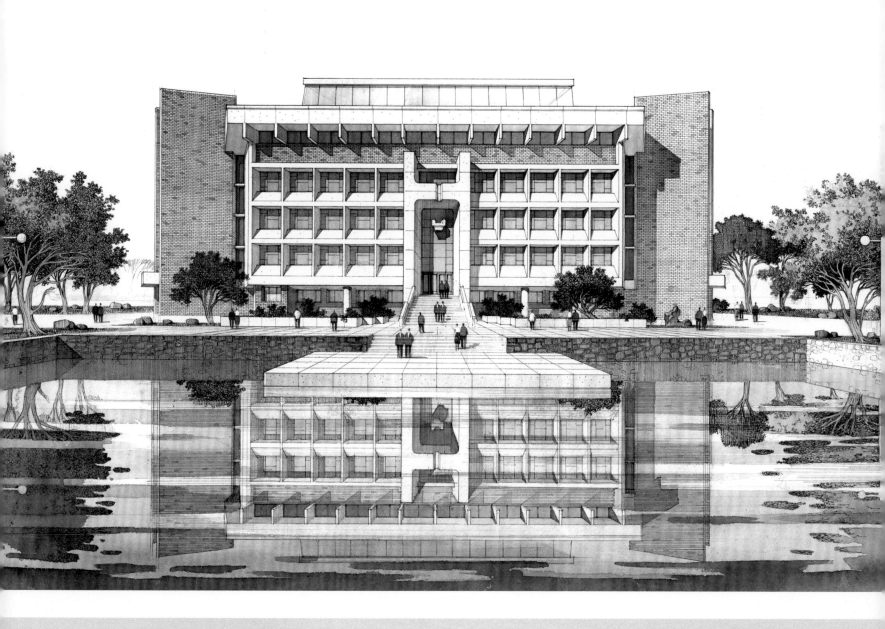

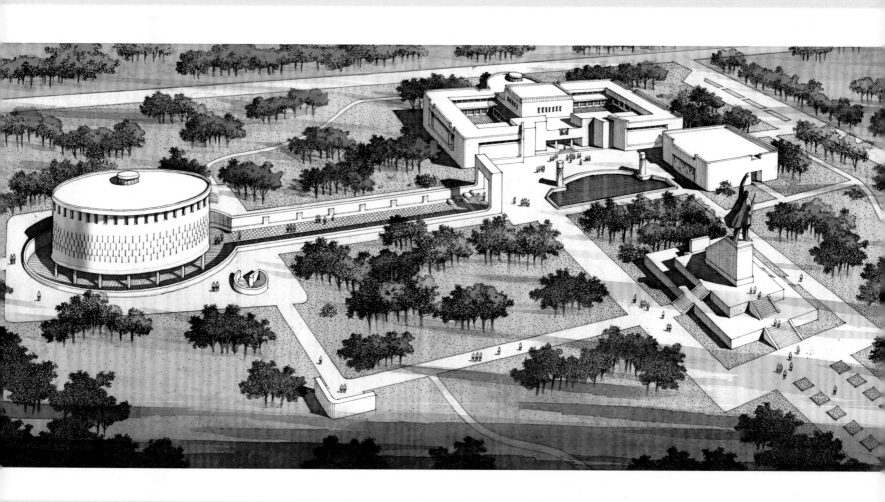

2

棕色篇

蓝色篇代表了一个时间段，继蓝色篇之后便是棕色篇。这个转型也与工具有关，即后来引进了德国生产的红环牌针管笔。这种笔最细可到 0.1 毫米，最粗可到 0.6 毫米，比意大利的产品又进了一步。此外，除黑色外，还有棕、红、绿、蓝等多种颜色的墨水。尤其是棕色（也就咖啡色），其作画效果比蓝色更好，于是我就弃蓝而取棕。这段时间延续得比较长，直至今日我还在用它来绘制各种建筑表现图。但是这种墨水不能用来渲染，所以，我只好用棕色的彩色铅笔来替代，即以棕色的线条和棕色的彩色铅笔相结合来绘制各种表现图。

BROWN

Those blue drawings belonged to a certain period of time which was followed by the brown ones. This transformation also occurred by tool. We applied ROTRING drawing pen and drawing ink which were made by Germany. The German pen is more advanced than the Italy pen because the thinnest pen is 0.1mm while the thickest one is 0.6mm. Besides, there's the color of red, blue, brown and green etc.. I like its brown most for the better drawing effect. I gave up the blues and used the brown color. It lasted a rather long time. Until now, I still use it to draw my presentation graphs. But this ink cannot be used to color. Then I have to use brown pencil and brown ROTRING ink to finish various drawings.

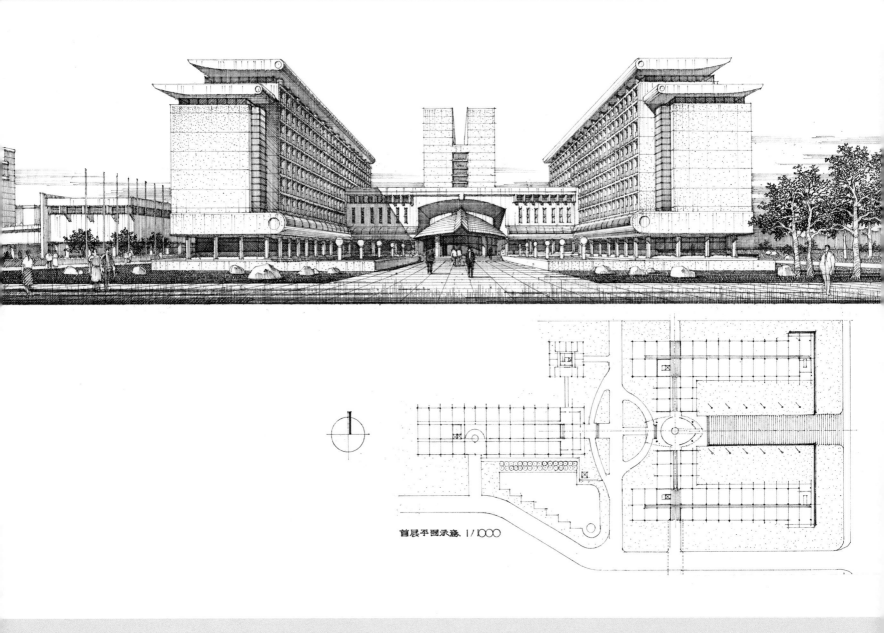

首层平面示意 1/1000

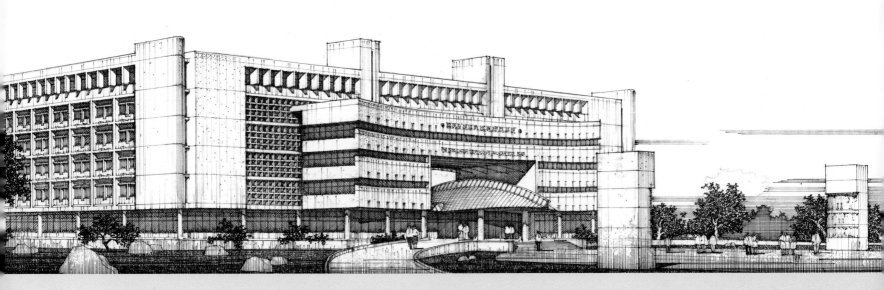

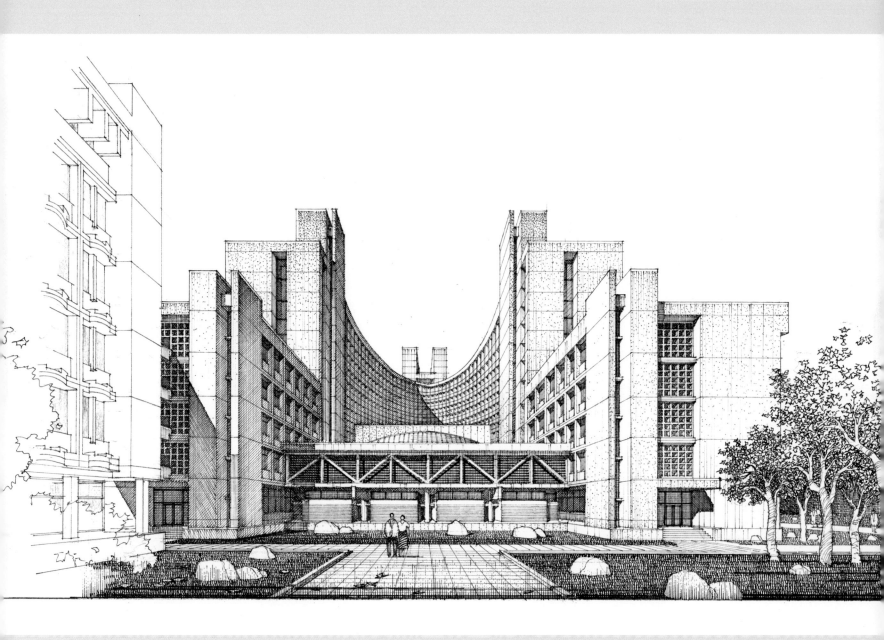

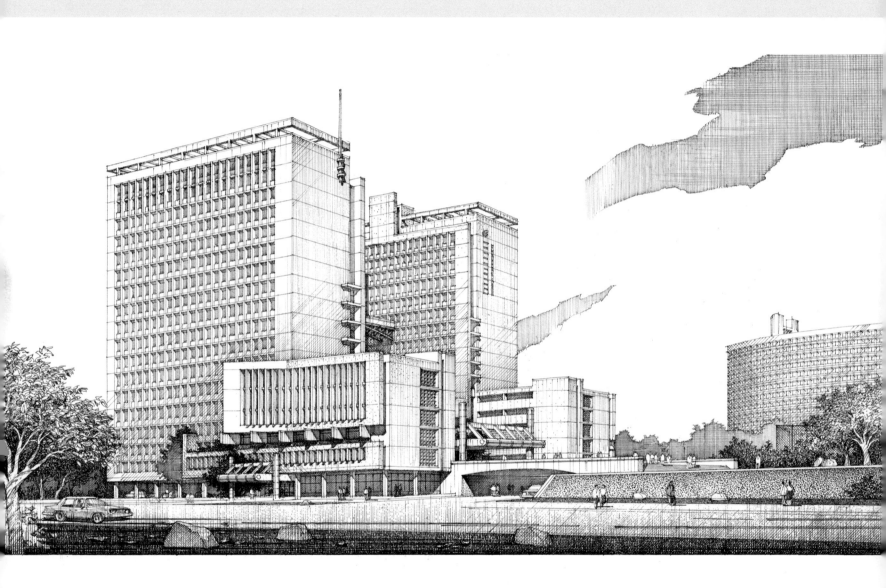

平面图 1/100

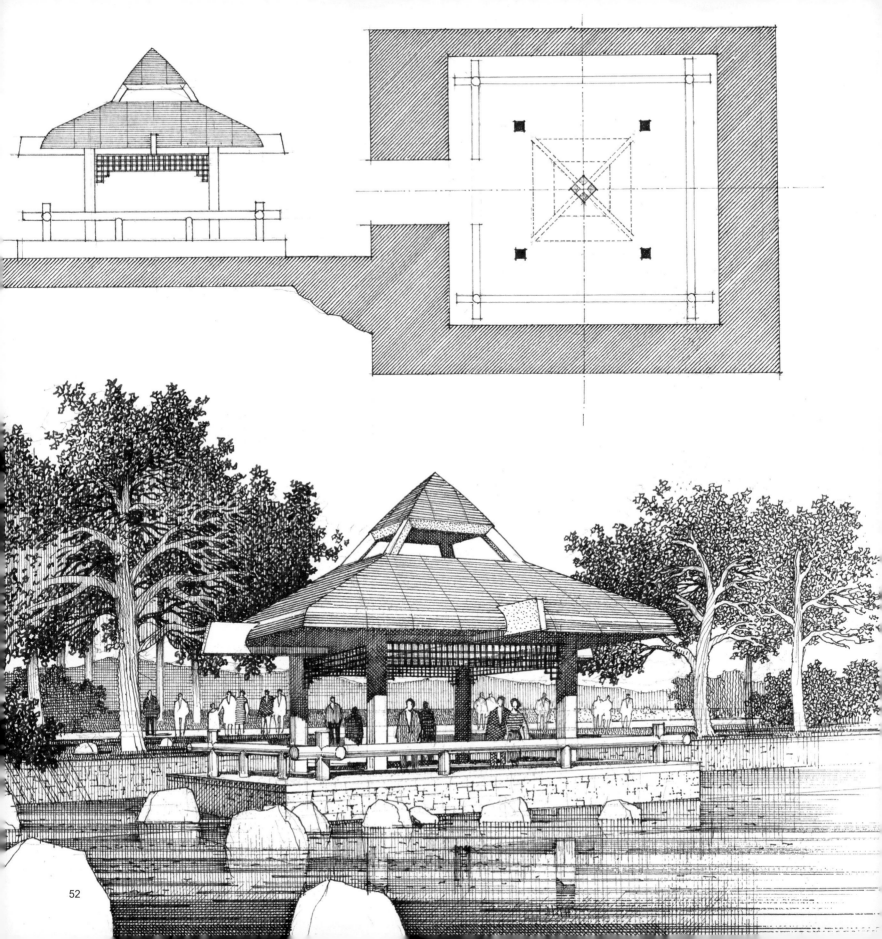

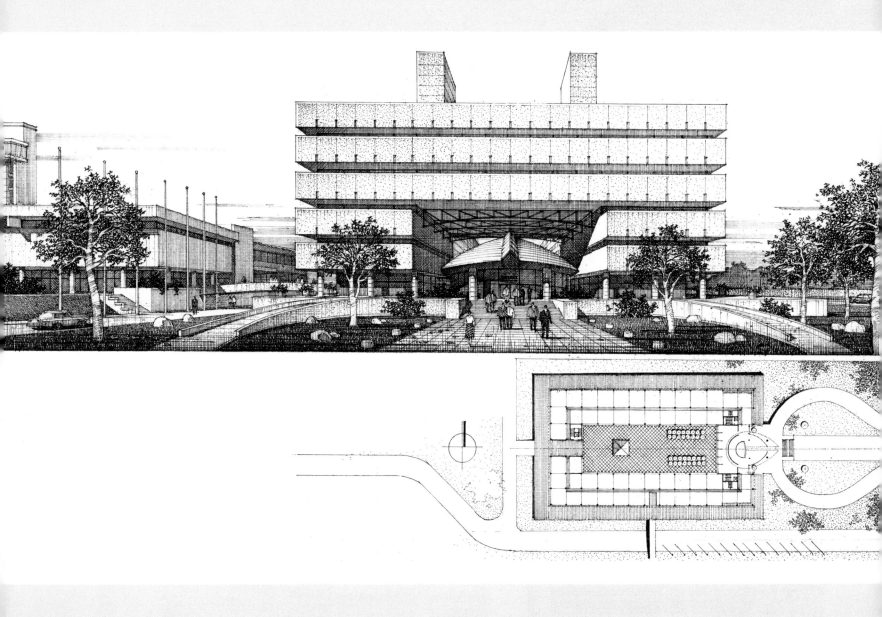

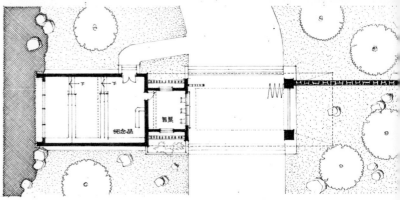

59

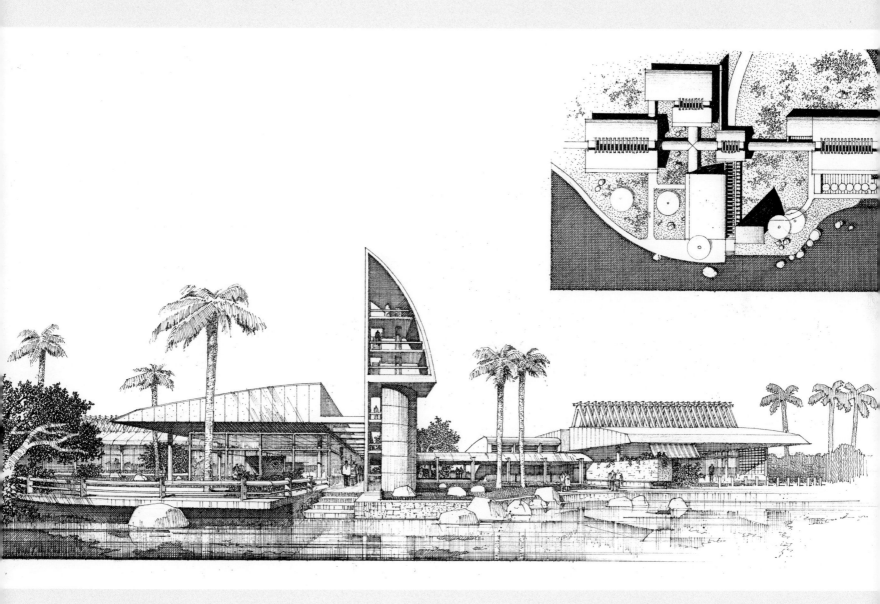

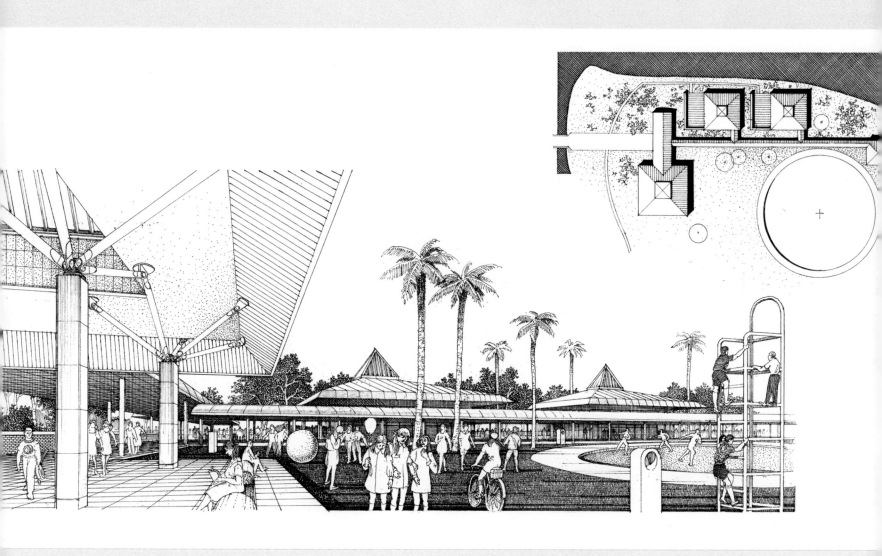

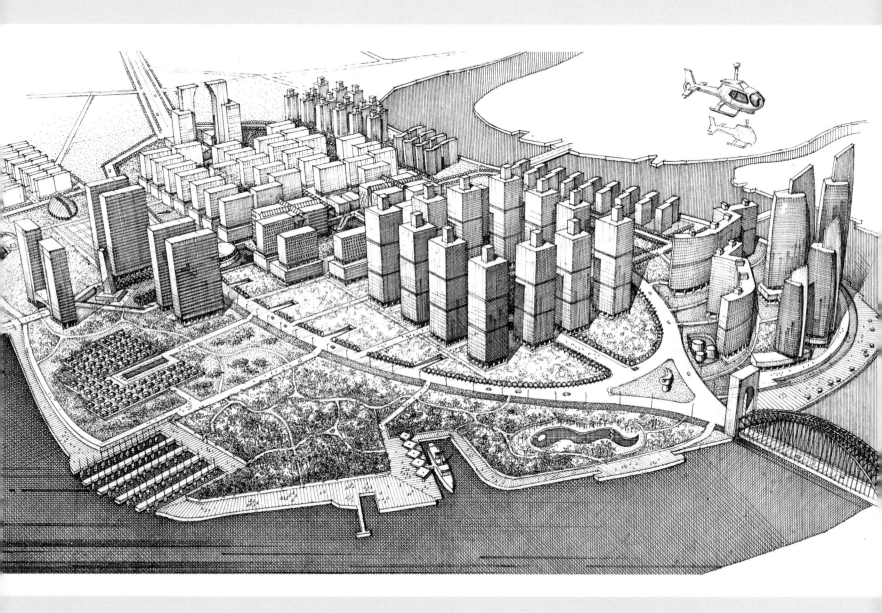

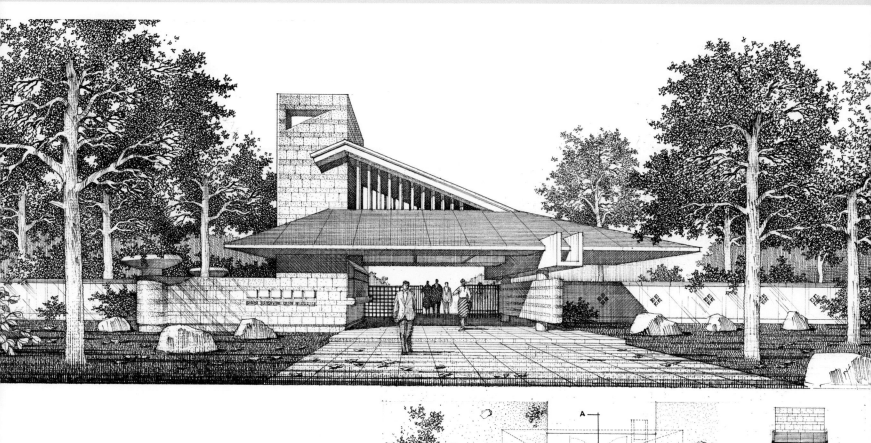
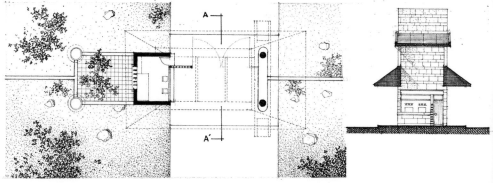

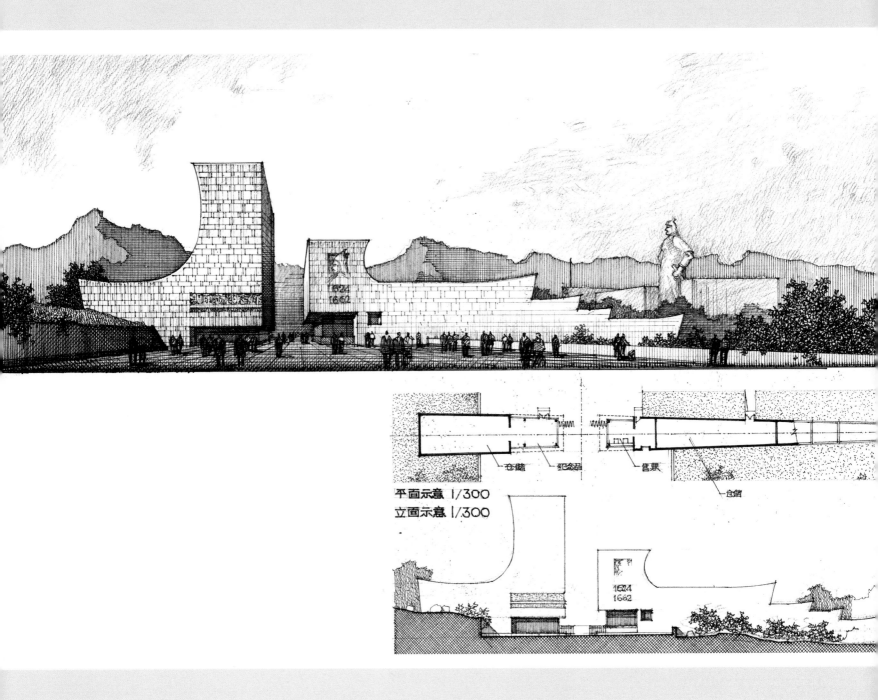

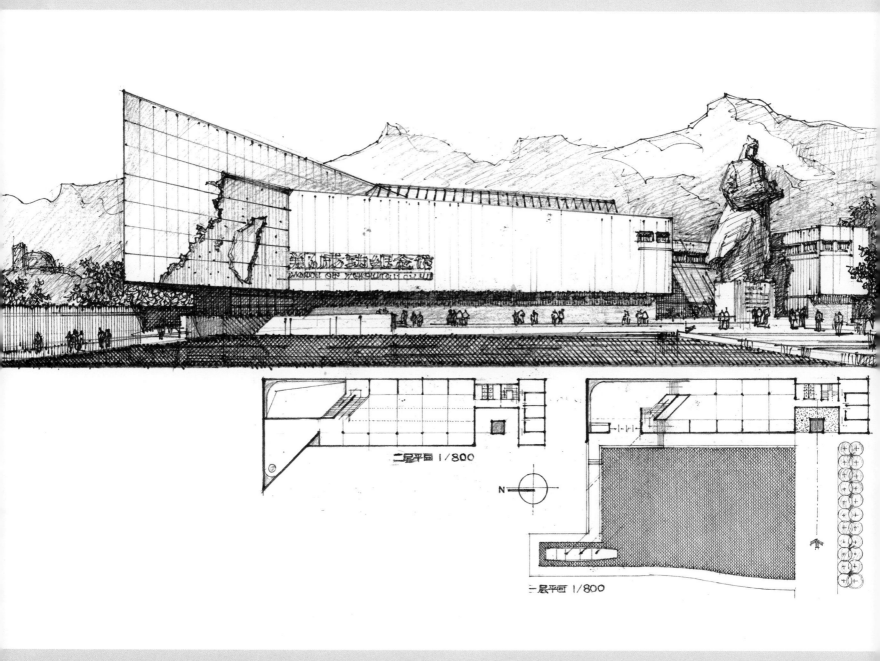

总平面图 1/1000
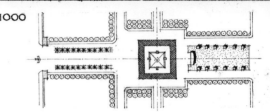

平面示意图 1/400

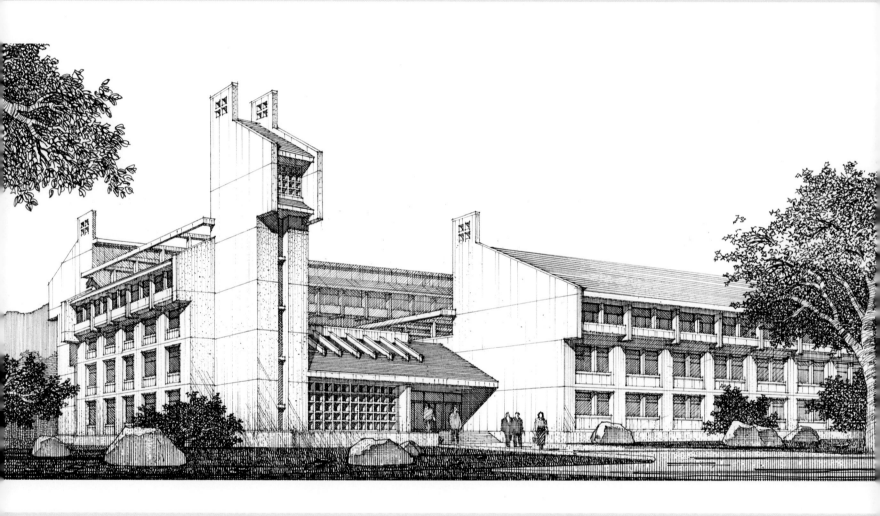

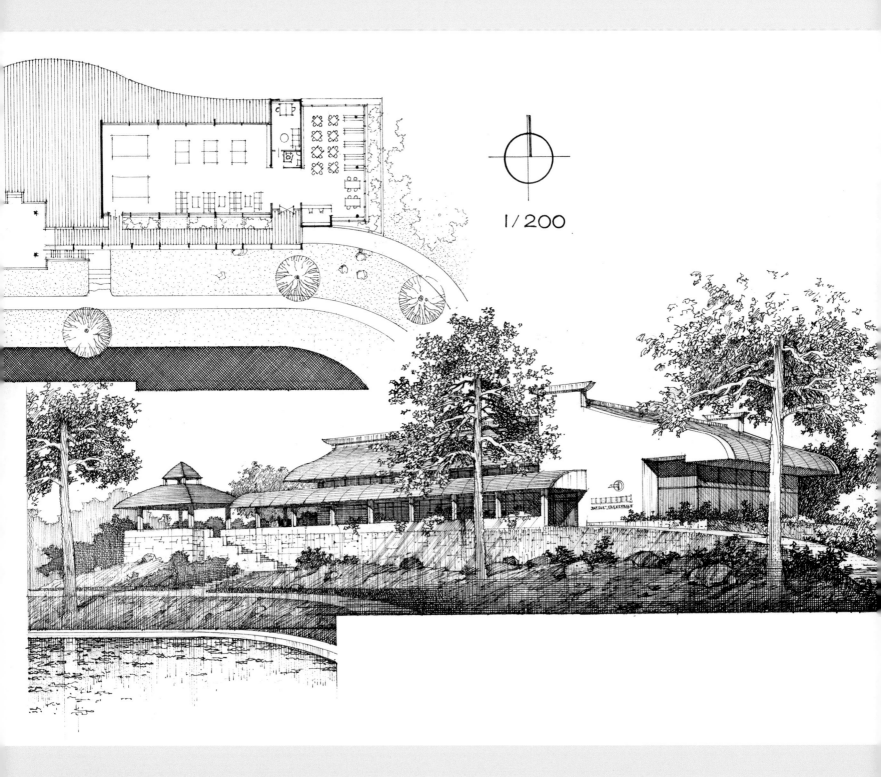

1/200

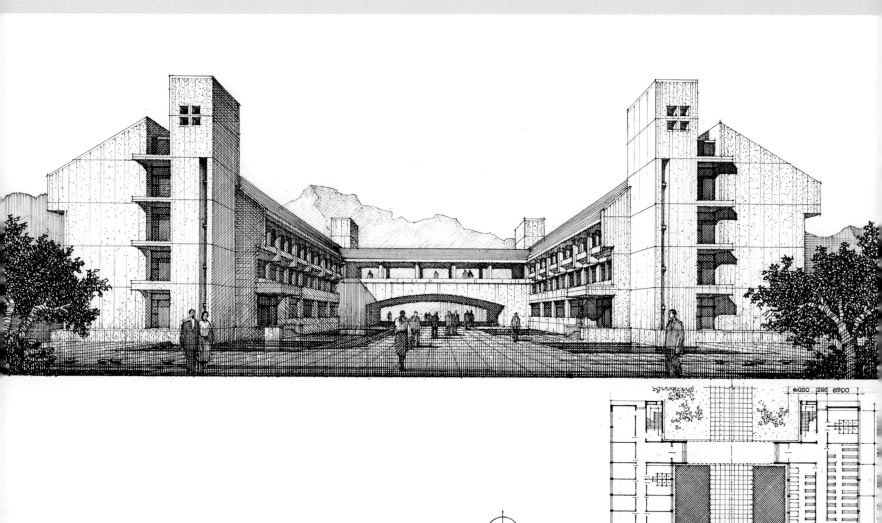

行政办公·图书馆
底层平面 1/400

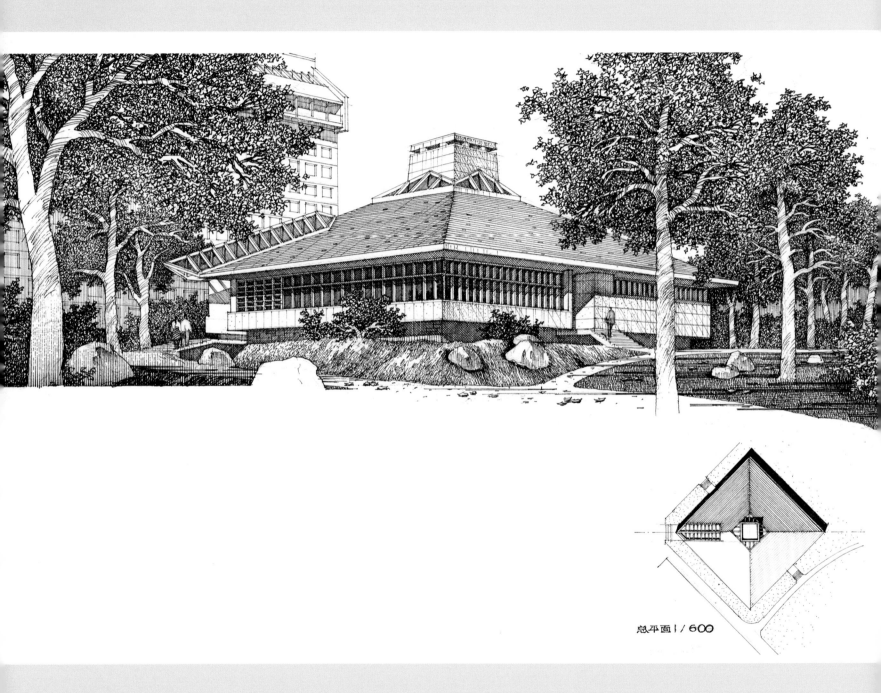

总平面1/600

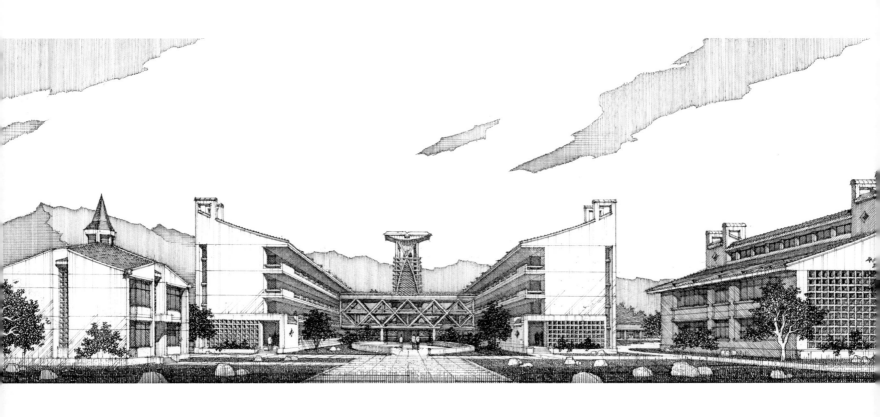

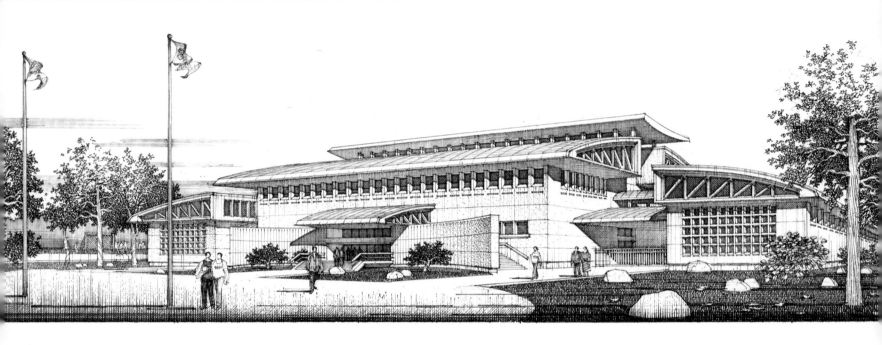
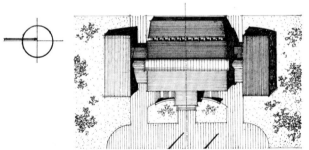

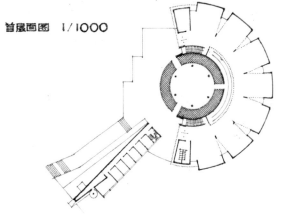

首层面图 1/1000

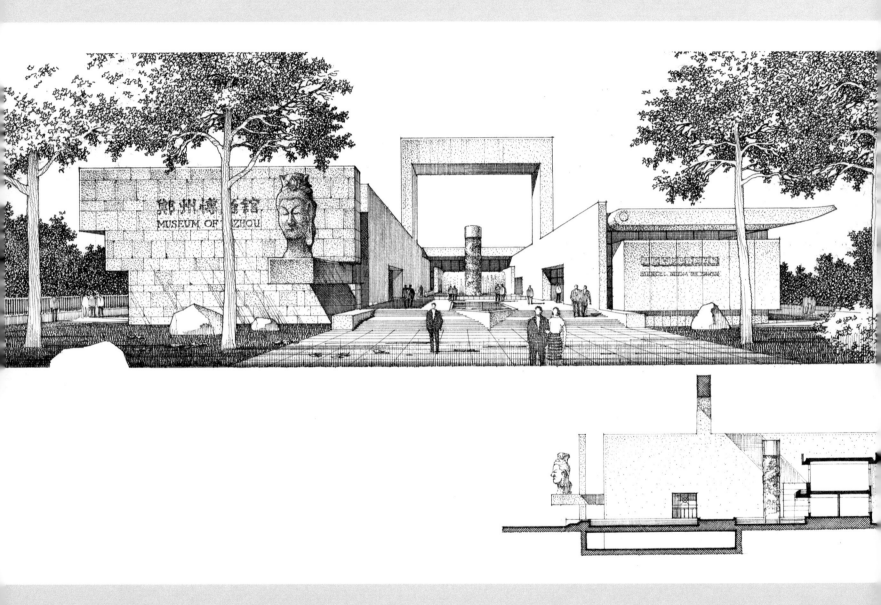

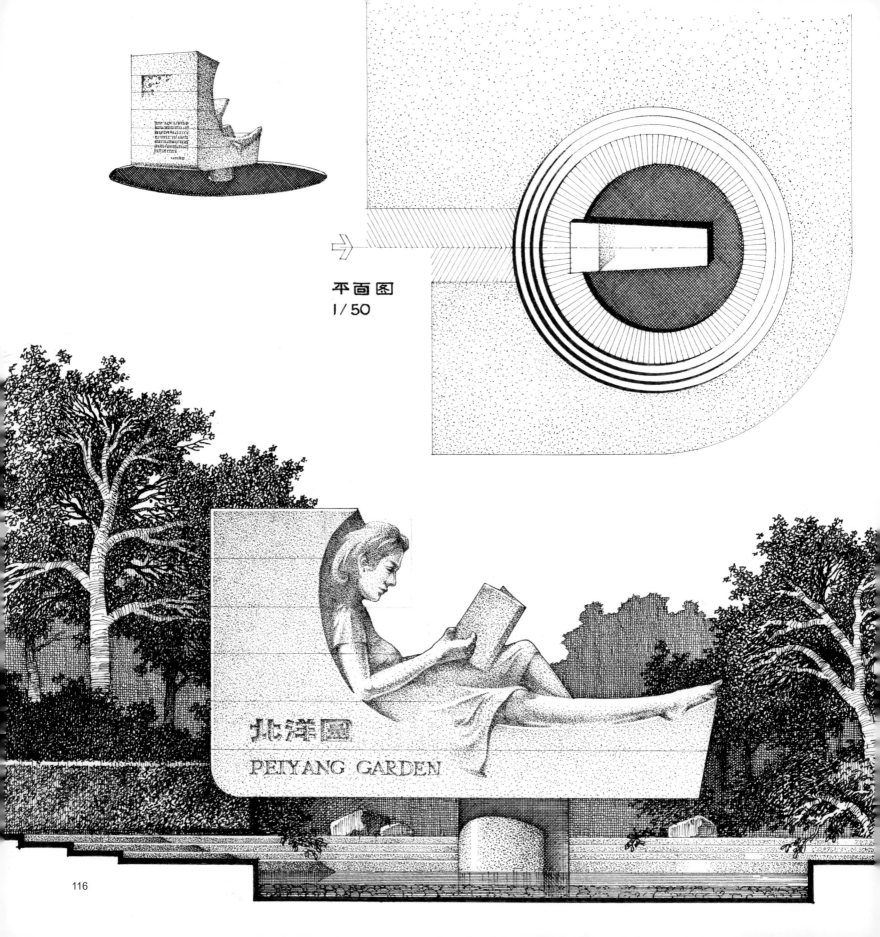

平面图
1/50

3

黑白篇

　　本集子所收入的黑白表现图是穿插在前两个时间段中所作的，它本身不构成一个独立的时间段。这主要是由两个原因造成的，其一是某些题材比较严肃，不适合用彩色线条来表现，其中，有的就直接画在灰色的卡片纸上，仅高光部分用了少量的白色，显得十分庄严。其二是由于轮廓线比较复杂，作完草图后，用半透明的硫酸纸蒙在草图上再细致地用铅笔完成轮廓线，这样就省事多了。这一部分表现图虽数量不多，但有自己的特色。

BLACK & WHITE

All the black and white drawings in this book were graphed between the former two periods of time. They could not form an individual period of time for the following two reasons: one reason is that some themes is too serious to present by colorful lines, for example, some drawings were drafted directly on the grey carton paper while just the color white used on the highlight part—to show its dignity; the other reason is about the complicated outlines. I had to cover the sulfate paper on the first sketch to finish the outlines carefully. There are no many drawings of black and white, but they all had their own specialty.

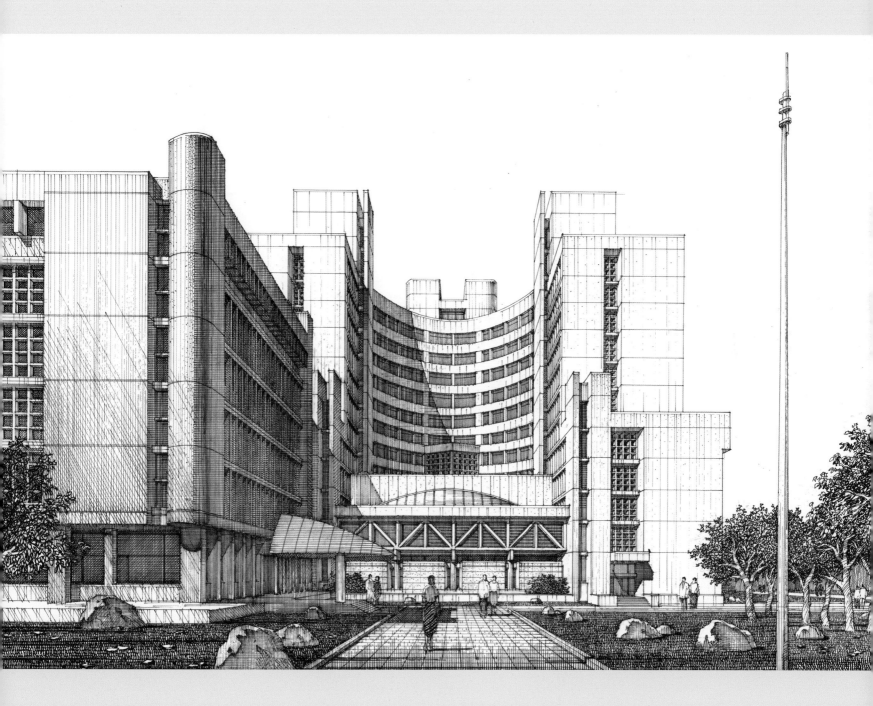

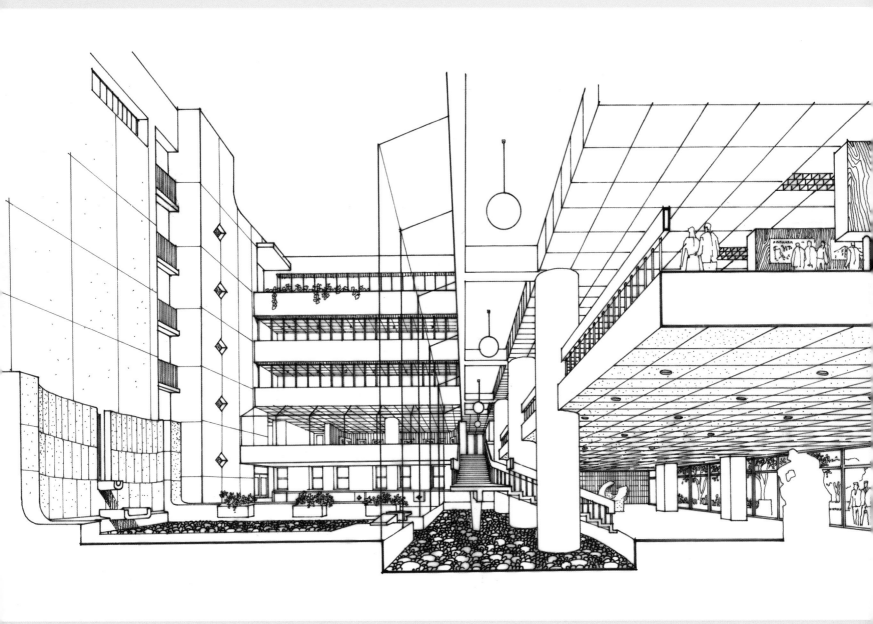

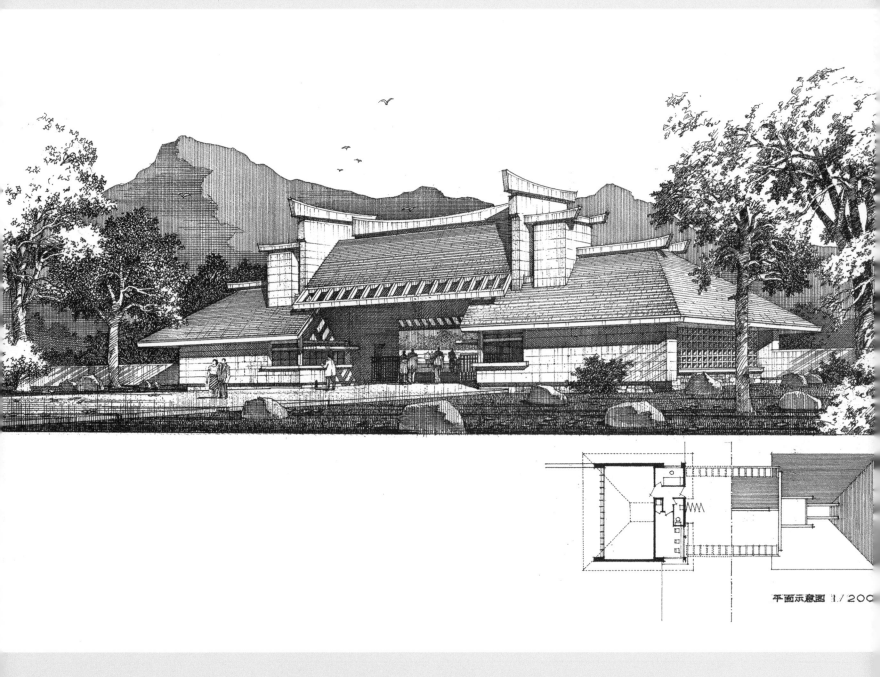

平面示意图 1/200

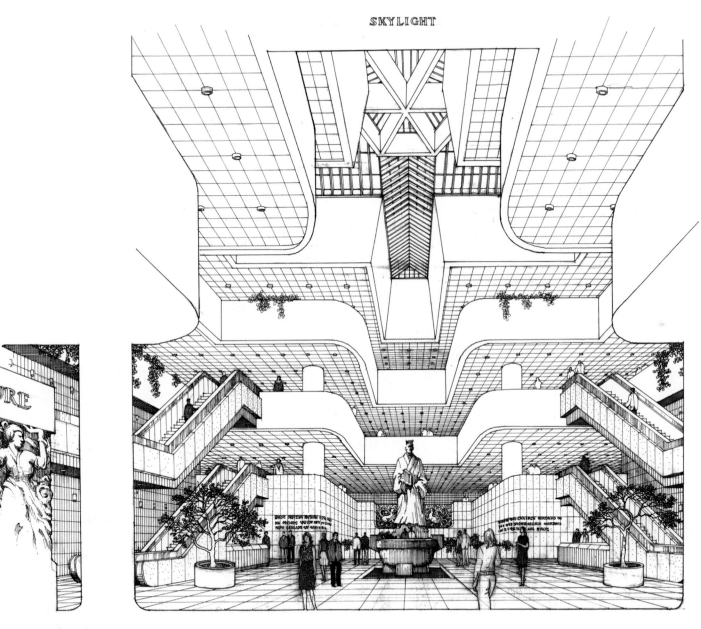

SKYLIGHT

SCULPTURAL DECORATION · QUOTATION · BACKGROUD DECORATION · QUOTATION · SCULPTURAL DECORATION
· CONFUCIUS STATUE
· INCENSE BURNER

总平面图

平面图 1/200

A-A' 剖面图 1/200

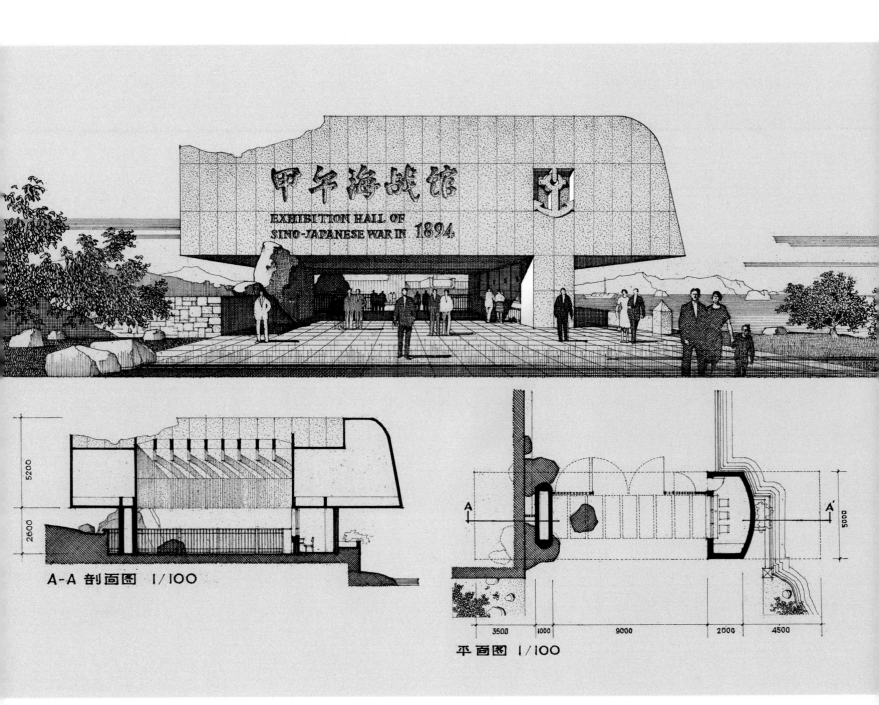

4

彩色篇

　　彩色表现图多半是应业主的要求而作的。在他们看来单色表现图根本算不上是表现图，只有着了颜色，方能真实地表现建成后的实际效果。这篇也可以分为两个时段，前一段为水彩渲染，很地道的传统方法，但是一种"湿"作业。首先要裱纸，渲染时还要一再地等着水干透后才能转入下一个程序。总之，很慢、很麻烦。后一段改为彩色铅笔着色，可以在原线条图上直接着色，也可以将原稿复印后在其上着色，这样，就方便多了。

COLORS

Most colorful drawings were drafted by the demand of owners. In their point of view, a single-colored drawing was not a real presentation graph. They thought that the real constructed effect could be only showed by a colorful graph. This period of time could be divided into two parts, the first part was done with watercolor and the second part was done with color pencils. Water coloring is a traditional way to draw, but it's a kind of wet work and it's slow and complicated. I had to mount the paper first and had to wait the paper dried when I colored the drawing. But the second part, color pencils brought me convenience. I could color the original lines directly or I could color the copied one. It's much convenient.

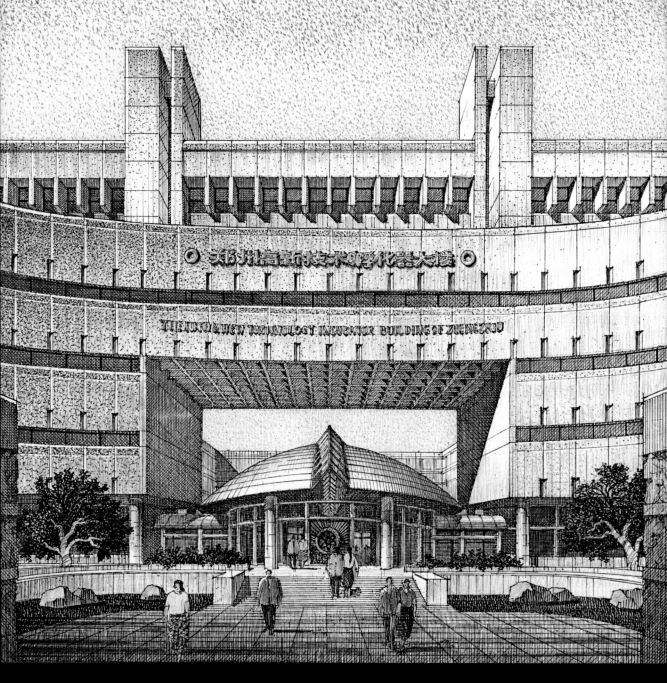

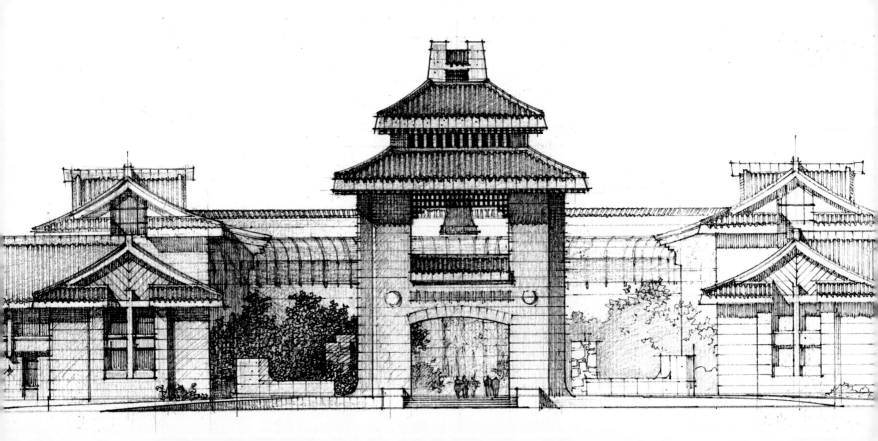

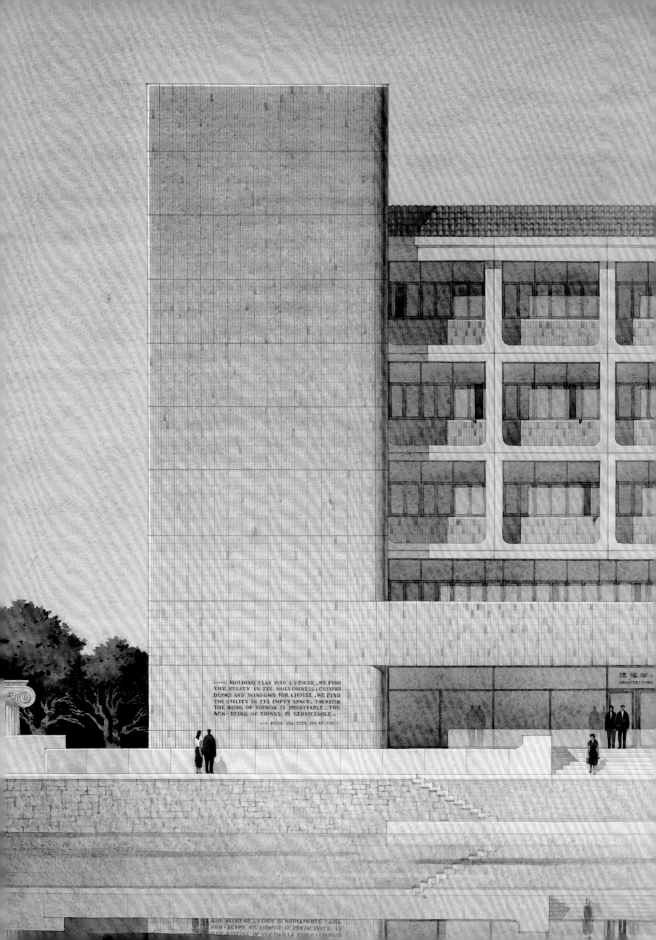

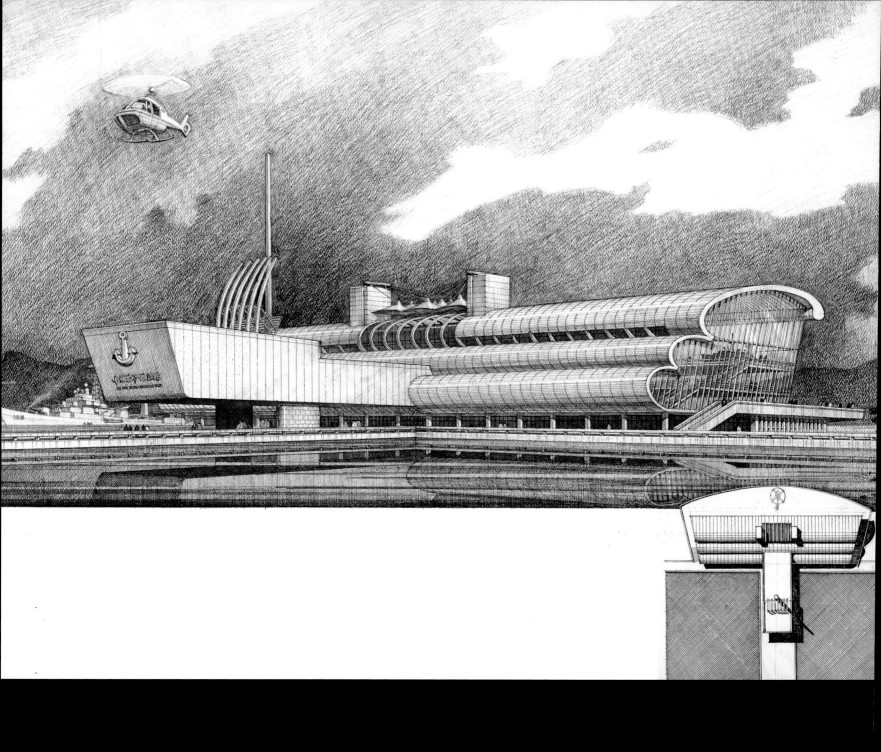

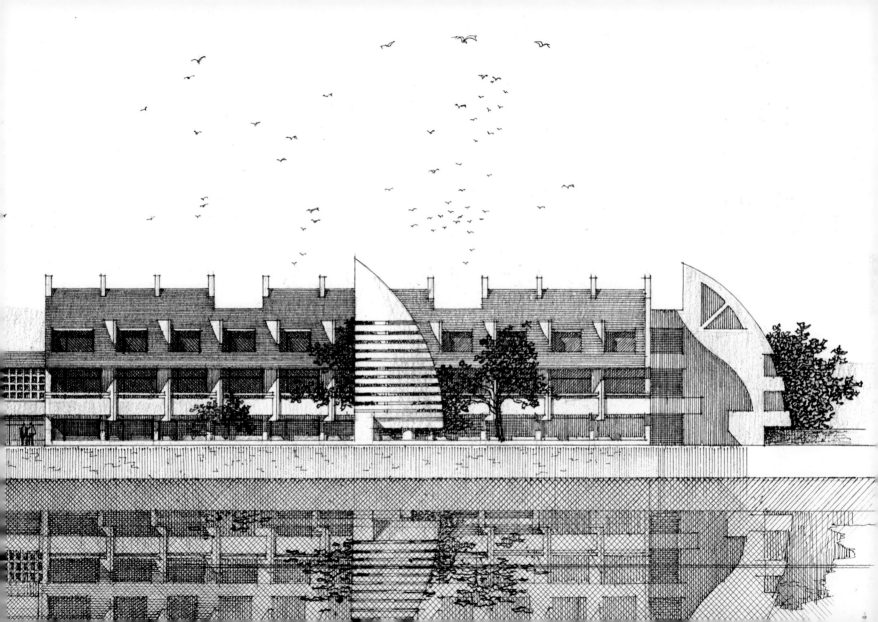

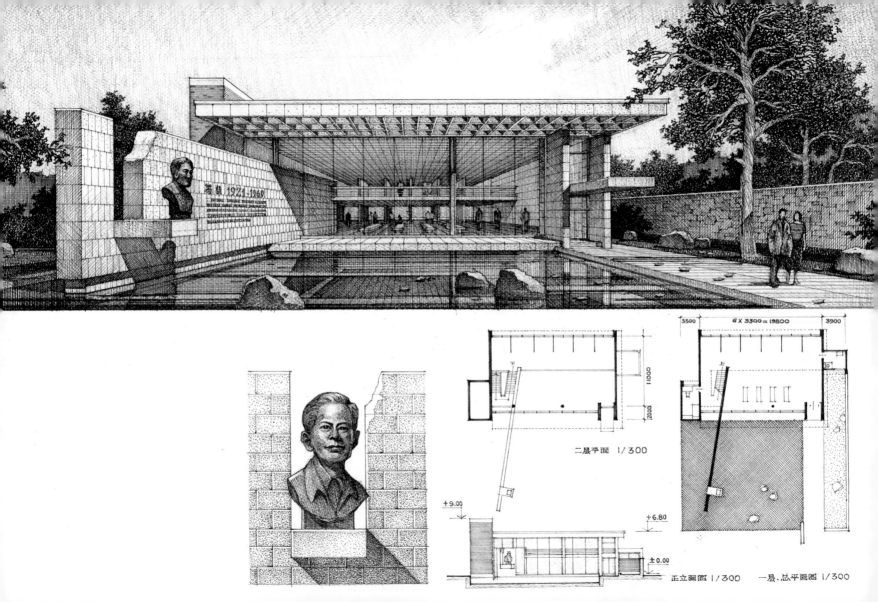

平面示意
1/500

观景厅堂
观景平台
观景挑台
视角盲区

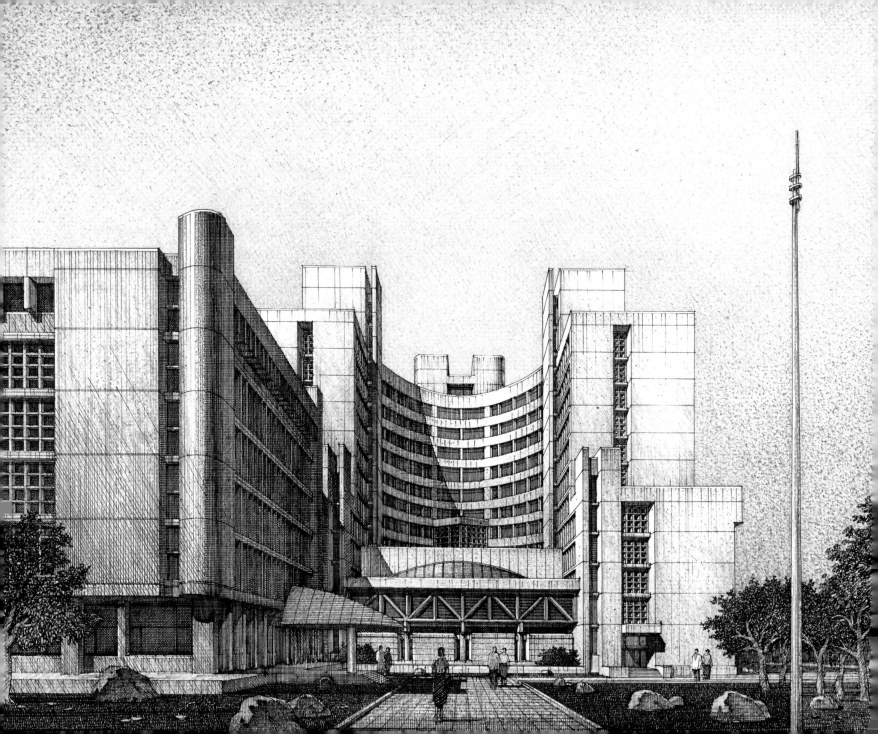

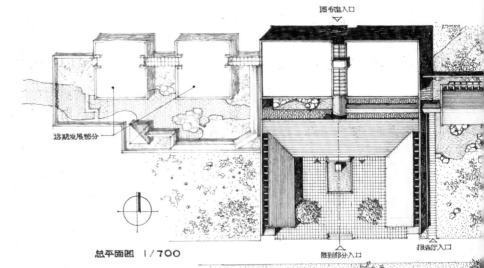

总平面图 1/700

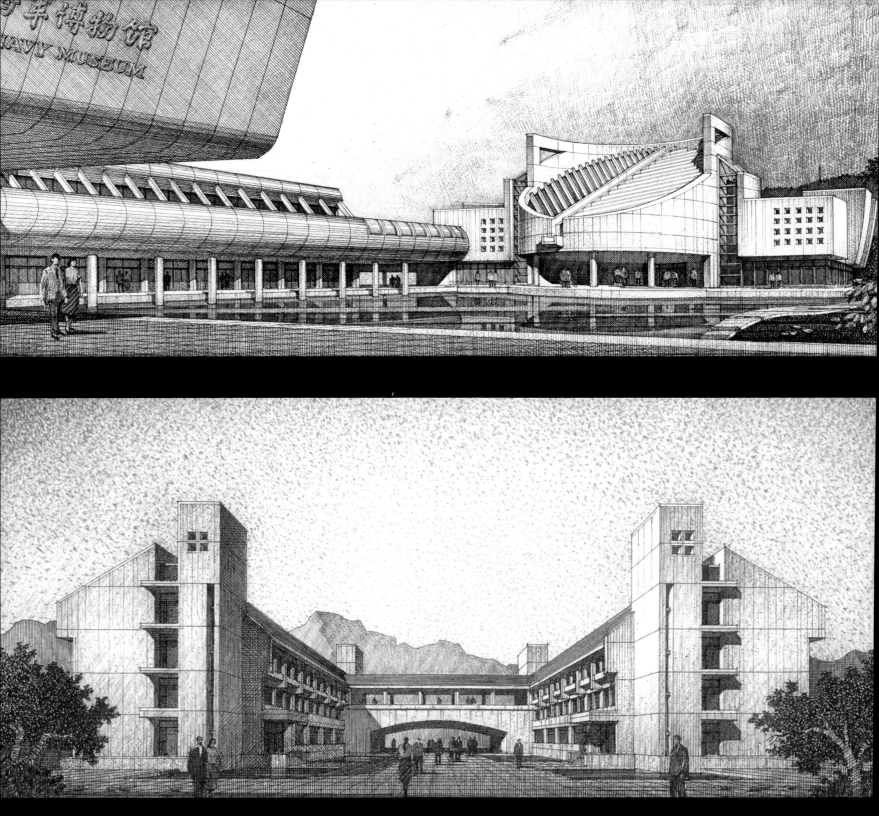

底层平面 1/400　楼层平面 1/400　总平面 1/800　A-A′剖面

校史陈列馆
1985.1.20.

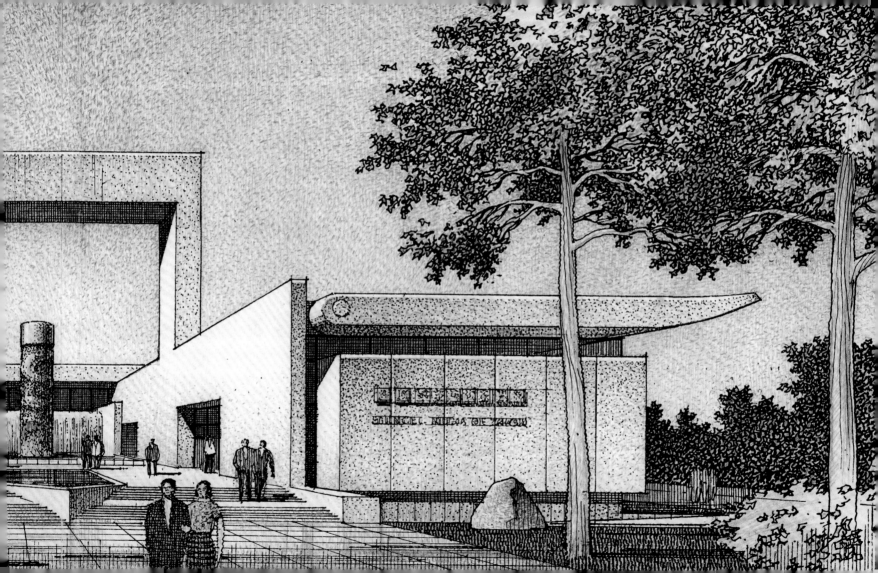

甲午海战馆

EXHIBITION HALL OF THE
SINO-JAPANESE WAR IN 1894

5

草图篇

对建筑师来讲，绘制草图的技巧至关重要，这几乎是电脑所无法替代的。草图所记录的是方案构思初期所闪现的一刹那间的想象，极朦胧、极粗犷，却为后来的方案发展提供一个很好的基础。这里所选的草图虽然有粗有细，但均为"原发"性的构思记录，没有任何刻意加工的痕迹。有的画在信笺的背面，有的画在外出宾馆时的小纸片上。多半是用钢笔绘制的，也有的是用彩色马克笔绘制的。只有少数几张用了尺子，那是在方案比较成熟时绘制的。

SKETCHES

The sketching skill is very important for architects. And it can't be replaced by computer. Sketches recorded the imagination occurred at the very beginning of the proposal design. It was vague, rough; but it's a basic thing for the further development of the design. The sketches in this book are all original ones without any alternation. Some of them were drawn on the back side of letter papers while some were drafted on the hotel memo papers. And they were drew with pen mostly or with colorful markers, only few of them was drafted with a ruler when the proposals were mature.

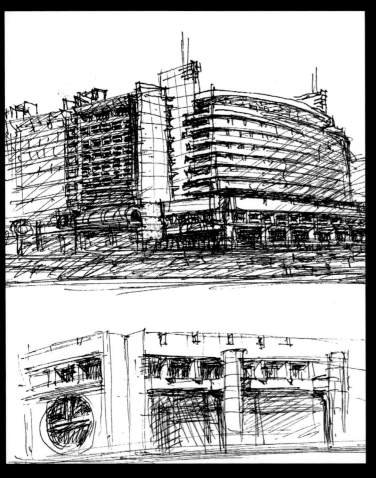

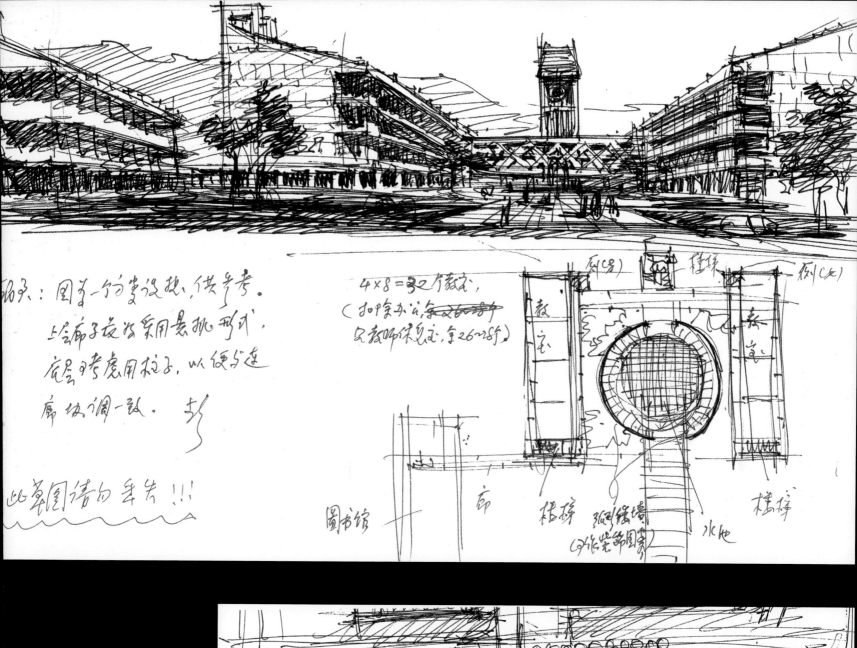

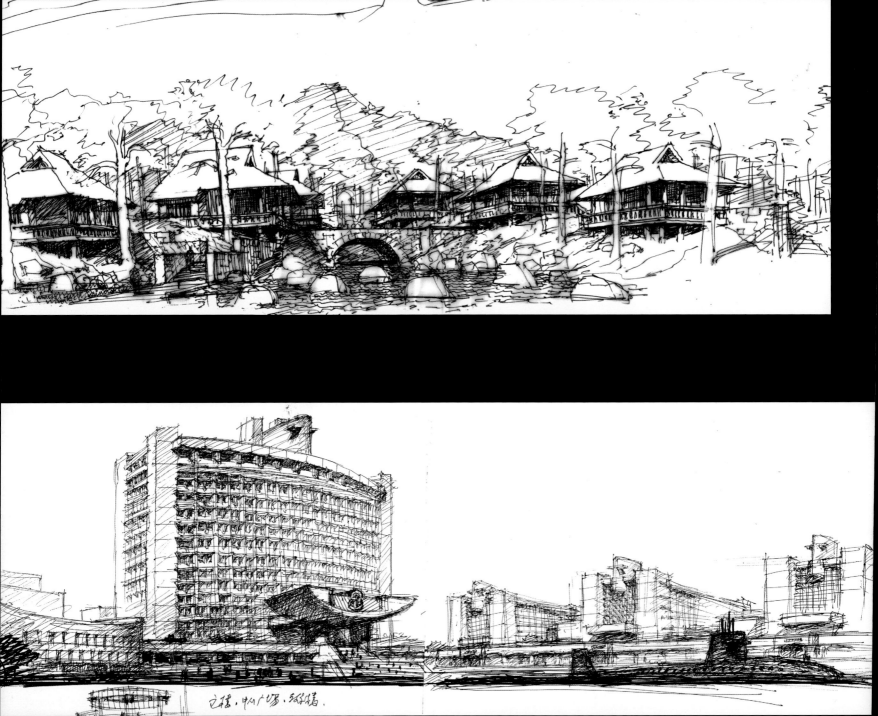

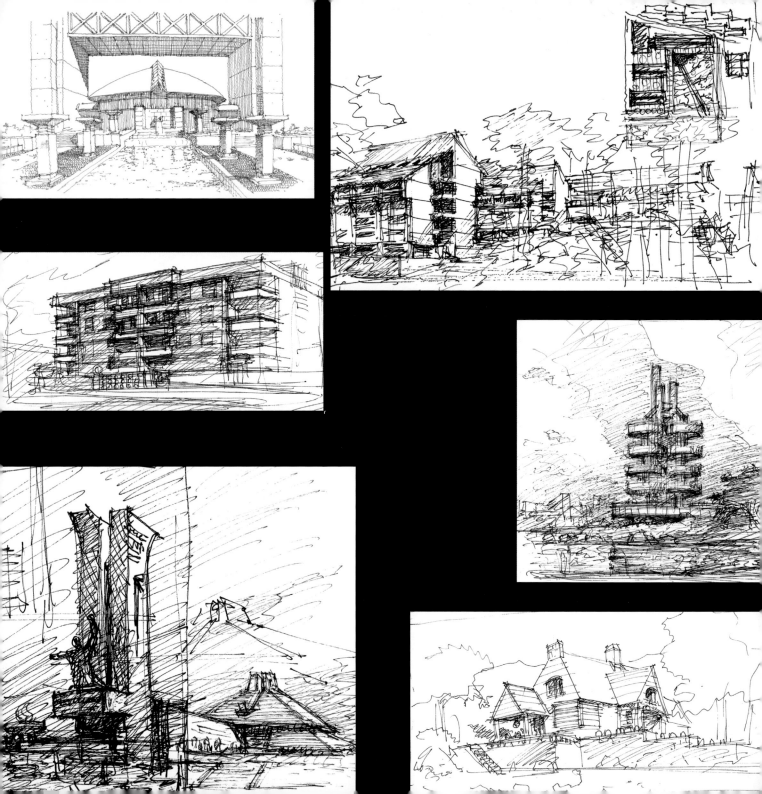

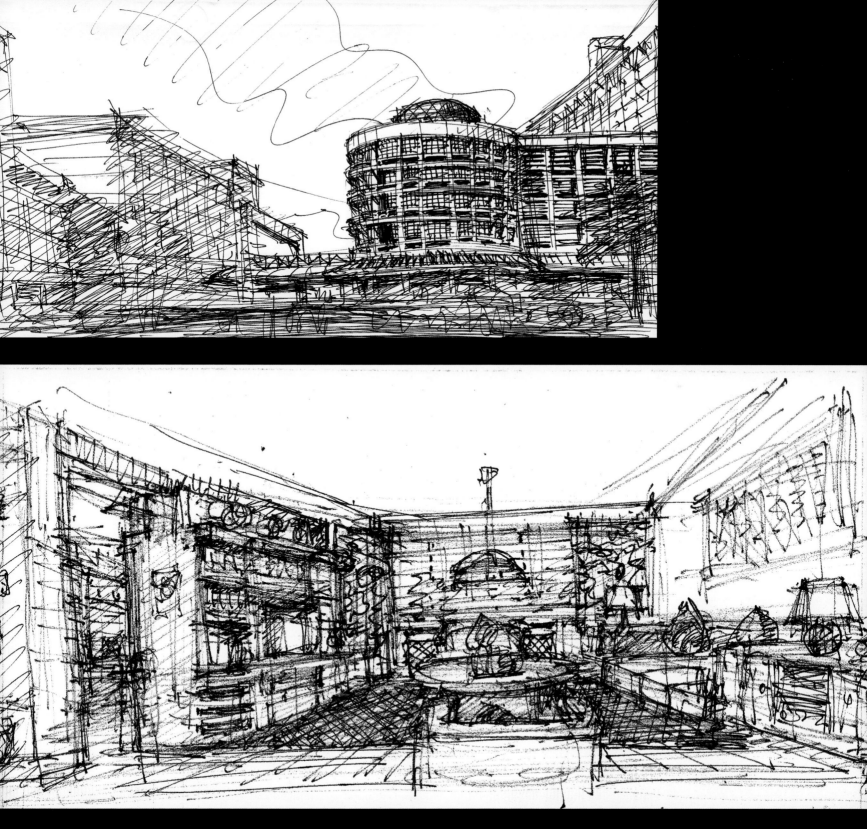

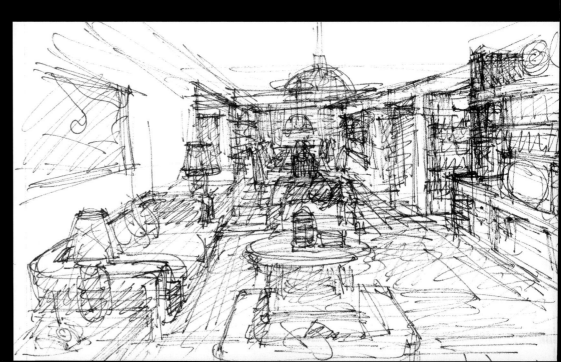

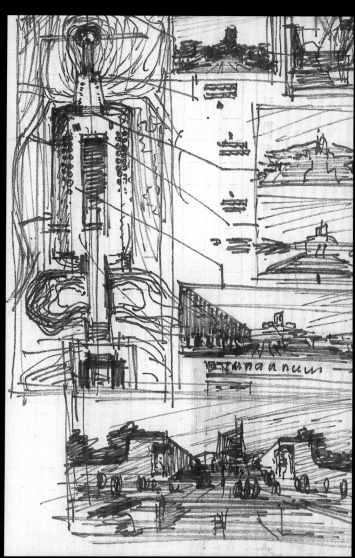

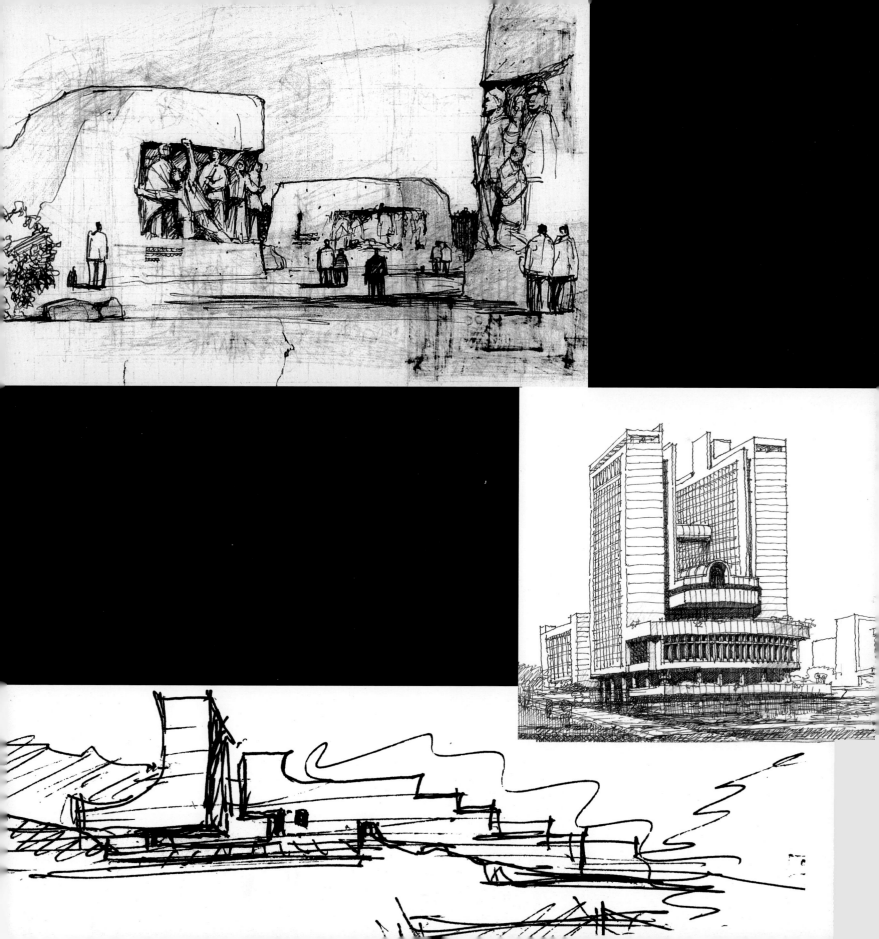

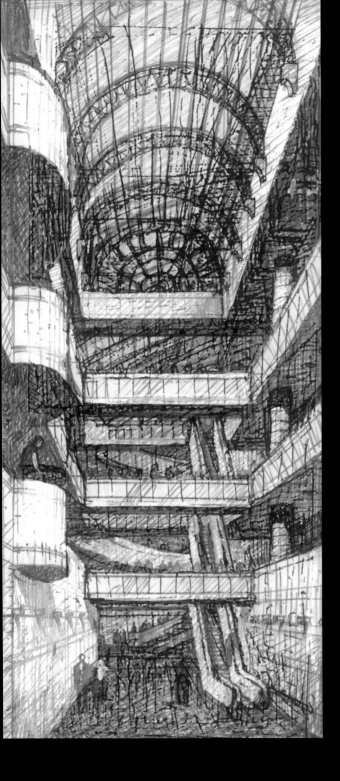

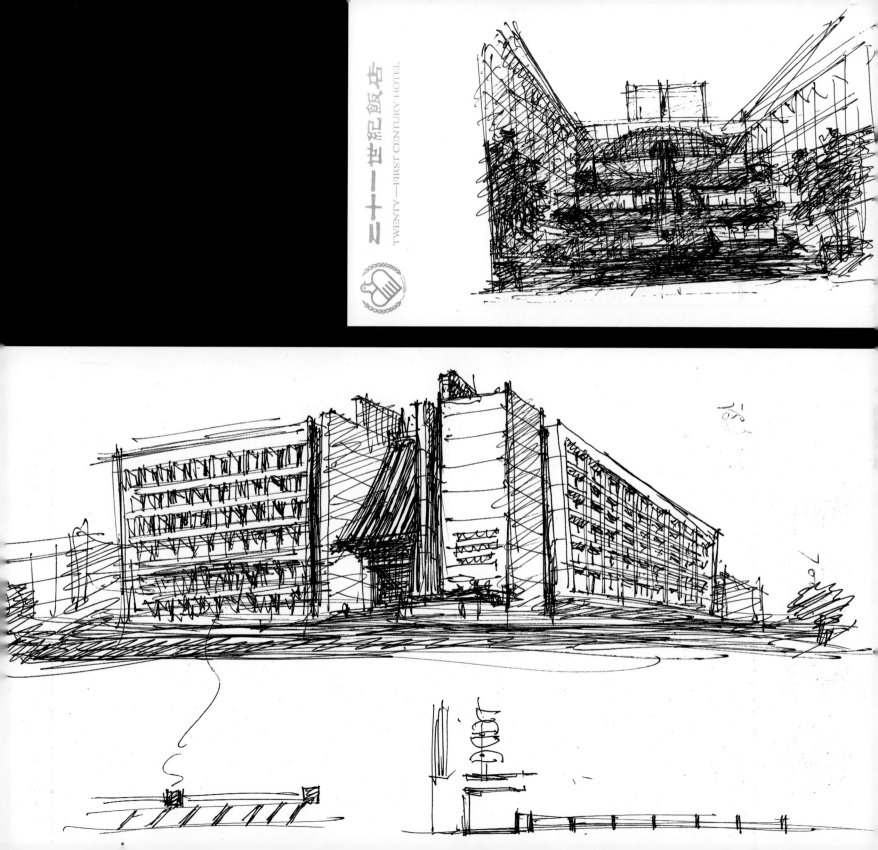

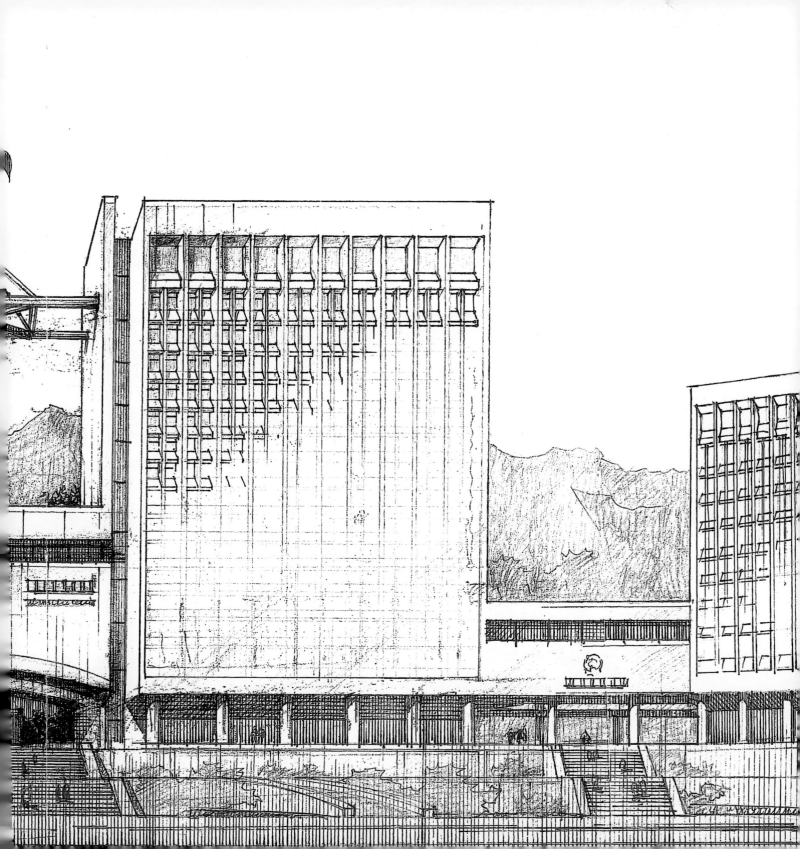

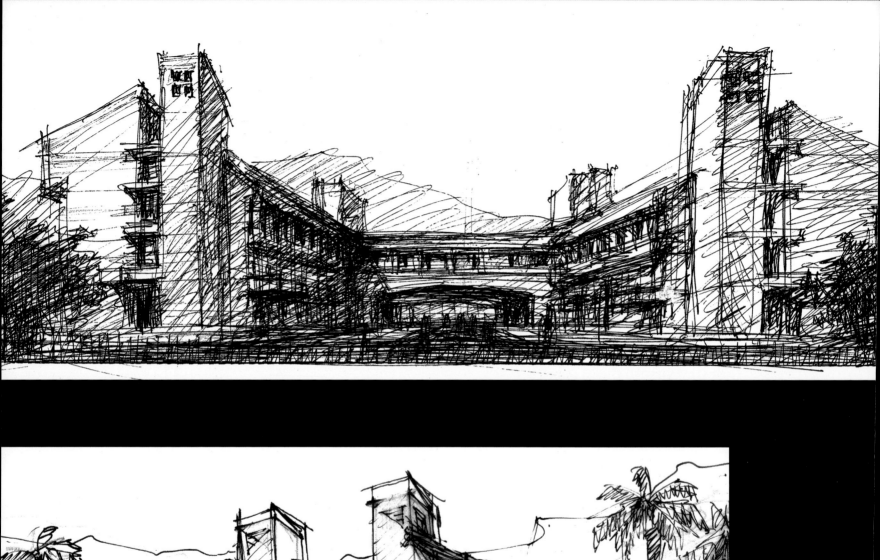

6

其他篇

　　这里所选的多属与建筑表现图无关的一些素描和人物肖像之类的作品。我的兴趣比较广泛，从小学时期就喜欢绘画，当时也没有老师指导，就对着照片临摹。学校虽然也开美术课，但是老师并没有认真教，我也没有认真学。我倒是对感兴趣的东西总想随手画画，于是便产生了兴趣。后来出版了一本《建筑师笔下的人物素描》，在这之后，又画了一些，这里便选了几幅，权当补缀。

MISCELLANEOUS

The drawings in this part include some pencil sketches and portraits which have nothing to do with architect. I liked drawing since I was in primary school. But there's no instructor at that time, so I just draw by imitating pictures. I didn't learn anything in class though we had fine arts lesson. However I just drew something that I interested in and developed the drawing interest from then. I published a book named CHARACTER SKETCHES BY AN ARCHITECT not long ago; and later I sketched some too. I put some of them in this book just for an addition.

临摹

临摹

临摹

临摹

临摹

四清时所住农家小院（河北赵县尉家庄）

7

后续篇

　　本篇主要包括：初版之后的某些新作，如天津大学新校区主入口校门、北洋纪念亭等；参加方案评审会时给某些入选方案的修改建议，以供原作者参考，如抗美援朝纪念馆扩、改建工程（未交付原作者）；某些没有按照邀请单位意见修改的方案，如福建晋江某标志性建筑；以及个别补遗的作品。在编排上没有分类插入原作的篇章，而是混杂集中在本篇之中。

SUPPLEMENT

This part includes: Mr. Peng's new works finished after the publication of the first edition, such as the main entrance of new campus of Tianjin University, Beiyang Memorial Pavilion, and so on; his suggestive amendments to many design proposals to be shortlisted, such as the renewal and extension projects of Korean War Memorial Hall (not delivered to the original author); some of his design proposals which are not amended according to the client's advice, such as the proposal for a landmark building in Jinjiang, Fujian; and other supplementary works. These works are not inserted into the original parts, just collected in this part.

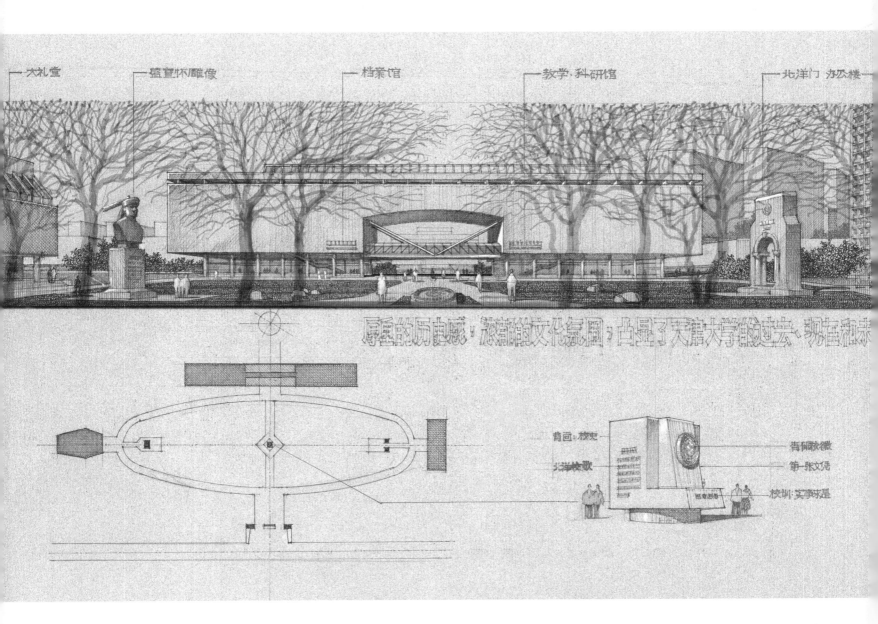

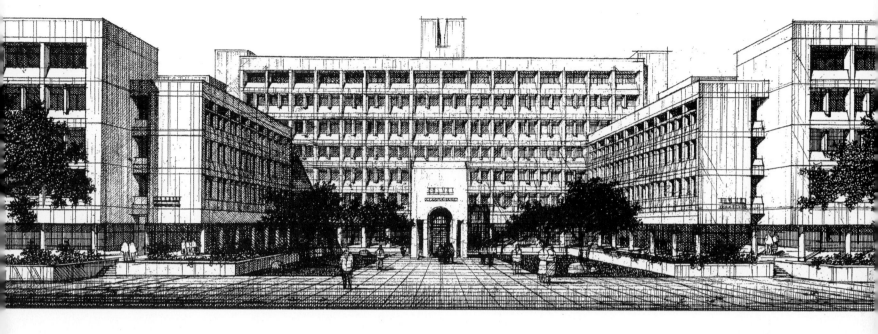

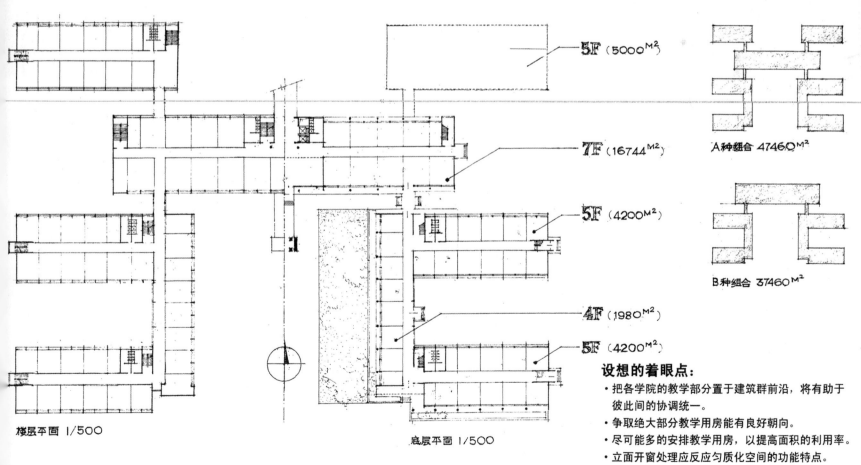

各学院教学楼部分平面设想

楼层平面 1/500　　底层平面 1/500

5F (5000 M²)
7F (16744 M²)
5F (4200 M²)
4F (1980 M²)
5F (4200 M²)

A种组合 47460 M²
B种组合 37460 M²

设想的着眼点：
- 把各学院的教学部分置于建筑群前沿，将有助于彼此间的协调统一。
- 争取绝大部分教学用房能有良好朝向。
- 尽可能多的安排教学用房，以提高面积的利用率。
- 立面开窗处理应反应匀质化空间的功能特点。
- 风格平实、朴素、典雅，但又具有适度的变化。
- 前院尺度亲切宜人。

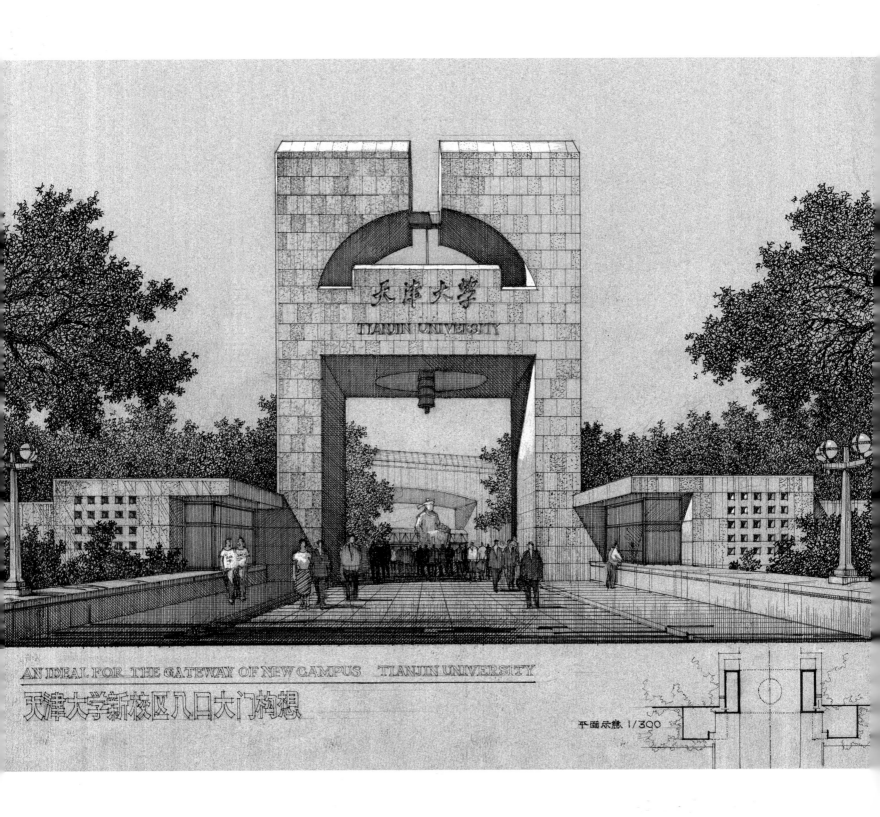

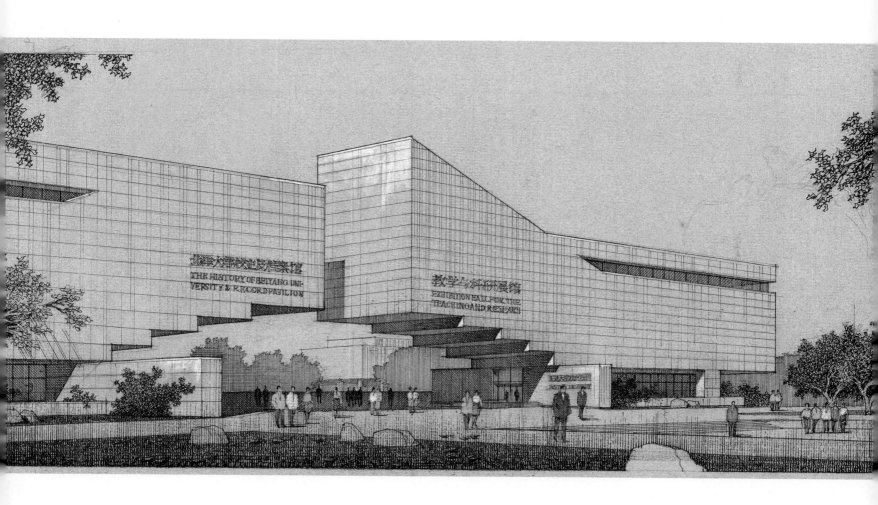

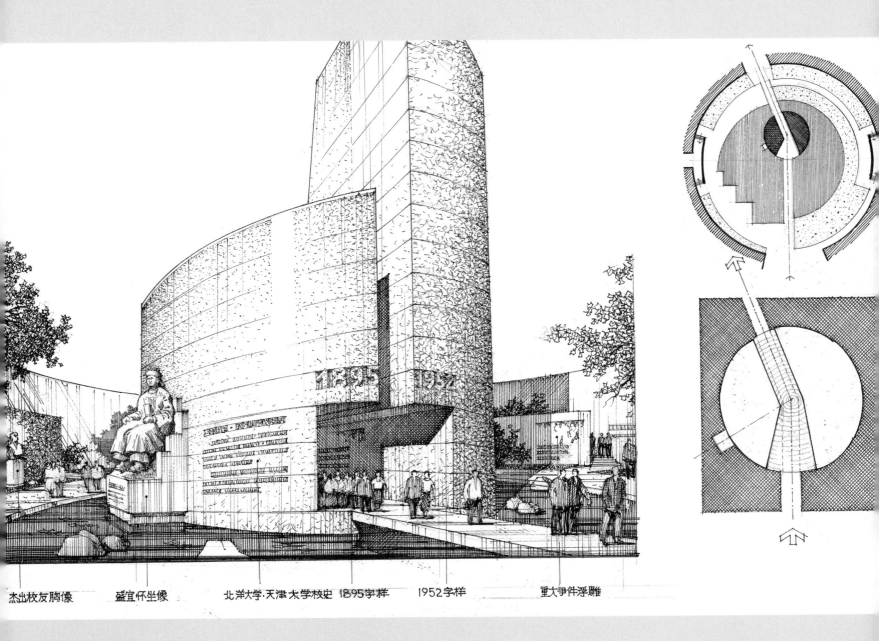

杰出校友胸像　　盛宣怀坐像　　北洋大学·天津大学校史 1895字样　　1952字样　　重大事件浮雕

平面示意 1/1000

天津大学新校区主入口
MAIN ENTRANCE OF NEW CAMPUS
TIANJIN UNIVERSITY   JUN·2016

凭眺阁构想 2013
AN IDEA FOR THE PAVILION
PINGTIAO

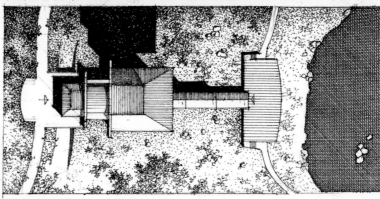

总平面示意 1/500

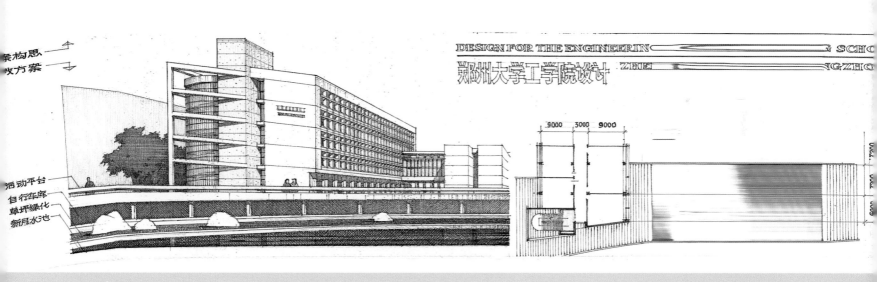

昌乐县文化馆方案设计

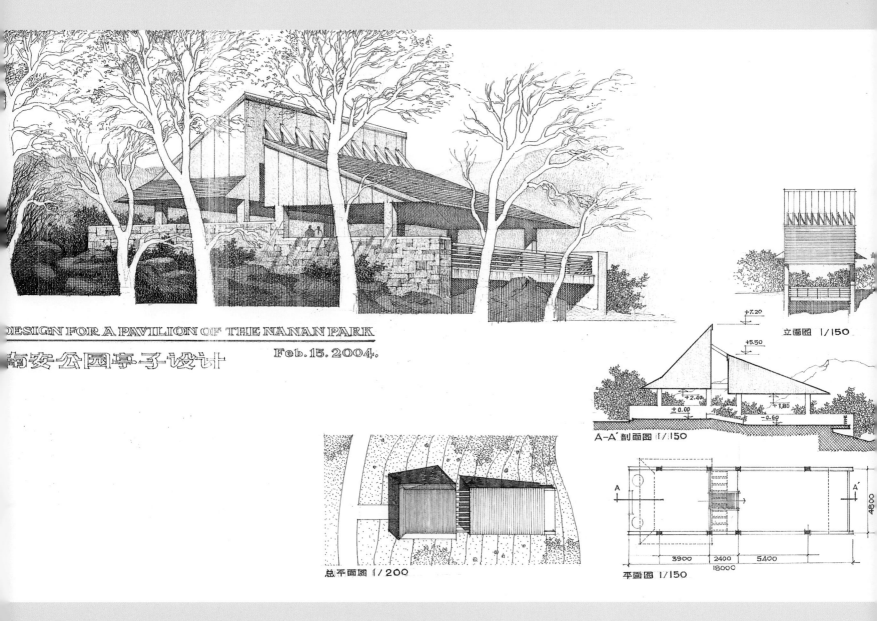

蜗居多年的筒子楼